O.A.W. DILKE is Profess
University of Leeds. He
at Stowe and King's Co
bridge, and has taught Classics at University College, Hull, the University of Glasgow, and The Ohio State University, and Latin at the University of Leeds. He is the author of several books.

ASPECTS OF GREEK AND ROMAN LIFE

★ ★ ★

GREEK AND ROMAN MAPS

O. A. W. Dilke

GREEK AND ROMAN MAPS

O. A. W. Dilke

with 62 illustrations

CORNELL UNIVERSITY PRESS
Ithaca, New York

This volume of Aspects of Greek and Roman Life
was conceived and commissioned under the general editorship
of the late Professor H. H. Scullard

First published 1985 by Cornell University Press

International Standard Book Number
0-8014-1801-1
Library of Congress Catalog Card Number
84-72221

Printed in Great Britain by
The Eastern Press Ltd., of London and Reading

CONTENTS

Appendices

LIST OF ILLUSTRATIONS

PLATES (between pages 96 and 97)

IN-TEXT LINE ILLUSTRATIONS

PREFACE

The part played by maps and plans in the history and civilization of Greece and Rome has not received the attention it deserves. Until recently the skills involved have been inadequately studied, and a very fragmentary and poorly balanced presentation has been made. Competence in map-making has too often been judged from the stylized interpretations of Greek world maps, the appearance of northern Britain in the co-ordinates and maps associated with Ptolemy's *Geography*, or the east–west elongation and north–south compression of the Peutinger Table, at least as it appears in the medieval manuscript. A greater acquaintance with ancient mapping on large and small scales suggests a very different assessment. Maps existed before Graeco-Roman times, and some large-scale plans had been carefully executed. But it was the theory of Greek scientists and the practice of Roman administrators and surveyors that ushered in a new era: in fact, if cartography may be defined as the science of map-making, the Greeks and Romans can be claimed as the first exponents of that science.

Many years ago the fascination of the illustrations in Roman land surveying manuals, especially where they were identifiable in relation to actual landscapes, started me on this research. I found that in histories of cartography there was no mention of the maps and plans in the Corpus Agrimensorum. Hence the need to write a book, which appeared as *The Roman Land Surveyors: an Introduction to the Agrimensores* (Newton Abbot, 1971). It is now out of print except in Italian translation, but a number of its illustrations and examples have been transferred to the present volume. Since then the Universities of Bologna and Pavia and societies as different as the Royal Institute of Chartered Surveyors, the Yorkshire Archaeological Society and the Association d'Etude des Civilisations Méditerranéennes have encouraged me to write and interpret for them. Papers delivered at the London Museum conference on Roman towns and at the Institut Français de Naples have contributed to the understanding of this and other aspects of Graeco-Roman mapping.

The long period of history involved and the changing roles of Greek cities and of Rome over many centuries, as well as the many types of map, presented a difficulty of arrangement. Chapters I–III, V and XI–XII adhere to chronological order, V dealing with the original Ptolemy while XI concerns the tradition of Ptolemaic maps. Chapters VI–X deal with themes rather than periods of mapping. As this book is intended to be on mapping of the earth, only an extremely brief outline is given of celestial cartography, where this is necessary for its implications. Non-mapping but training aspects of the work of land surveyors have been included, and also descriptions marginal to mapping such as geographical writings, itineraries and *periploi*. Greek names have been spelt in Latin form where they are familiar. The aim has been to provide for the interests not only of classicists and historians of cartography but of many other readers.

Special thanks are due to my wife, who has helped my work throughout. The Leverhulme Trust awarded me an Emeritus Fellowship providing for travel and research to complete writings which are now gathered in this book and also in Vol. 1 of *History of Cartography*, edited by Brian Harley and David Woodward for Chicago University Press. I wish particularly to acknowledge access to the contributions to that series of Professors Germaine Aujac and J. Babicz, Dr A. R. Millard, Professor A. F. Shore and Dr Catherine Delano Smith. Acknowledgments are likewise due to Dr Helen Wallis and others at the Map Room of the British Library, and to the Special Collections of the Brotherton Library and the Audio-Visual Service, University of Leeds. The challenge of the Schools Museum Service, Wakefield, which had made models of Roman surveying instruments, to prove with students that these worked, enabled us to see that the Romans had easy and accurate methods for making large-scale plans. The John Carter Brown Library, Brown University, Providence, Rhode Island, where at present I have a research fellowship, has by its rich collection enabled me to appreciate better the rediscovery of Ptolemy's *Geography* and its effects on the later period.

CHAPTER I

THE PREDECESSORS

MESOPOTAMIA

It would seem probable that the concept of maps and plans developed independently in different areas of the world, among both literate and pre-literate peoples. It is now known that China and the Graeco-Roman world were producing maps at about the same period. Some of these had at least one feature in common, the use of co-ordinates, but there are many differences which suggest independent development. The Çatal Hüyük 'town plan' (p. 102 below) is not likely to have influenced Greek plans. In the case, however, of Mesopotamia and Egypt, and perhaps also of Persia, there is some reason to think that map-making may have influenced Greek theory and practice. Greek map-making started at Miletus and other places in western Asia Minor; these Ionian scientists, philosophers and historians had strong links with Egypt and Mesopotamia and needed maps to support them in their plans for a successful revolt from the control of the Persian Empire.

In Mesopotamia the invention by the Sumerians of cuneiform writing in the fourth millennium BC paved the way for the production of maps.[1] Cuneiform tablets of the period between 2500 and 2200 BC include long lists of place-names, rivers and mountains. These may have been used for teaching or perhaps military purposes rather than picked out of a map in the manner of the Ravenna Cosmographer of some 3000 years later, but whether they suggest the existence of maps is uncertain. The fact that King Sargon of Akkad was making military expeditions westwards from about 2330 BC would account for the inclusion of places as far west as the Mediterranean. Later we encounter itineraries, referring either to military or to trading expeditions. They do not go so far as to record distances, but they do mention the number of nights spent at each place, and sometimes include notes or drawings of localities passed through. As in Greek and Roman inscriptions, some documents record the boundaries of countries or cities.

The earliest large-scale plan from Mesopotamia with an indication of scale is one incised on the robe across the lap of King Gudea of Lagash, Sumer (*c.* 2100 BC), now in the Louvre, Paris.[2] This appears only on one stone statue out of several of Gudea extant. There are traces, on a broken corner, of a scale-bar; this ties up with later plans drawn to scale. The plan seems to represent thick walls with a complicated system of bastions, and with six rather narrow entrances. Although it has been called a temple or palace-hall, it looks more like a fortress. All corners are right angles, but it is obviously accommodated to the contours of the terrain.

Large-scale plans are found on clay tablets from the time of Sargon onwards. Incised lines could indicate rivers, canals, walls or streets, so that some ambiguity is inevitable. Extant examples vary considerably in cartographic exactness, the most precise having parallel lines carefully drawn and recording measurements of rooms. The most interesting is a plan of Nippur, the Sumerian religious city, of *c.* 1000 BC.[3] It shows the Euphrates, two canals, the main temple, two enclosures, and the city wall, with seven named gates. Measurements given for certain buildings are thought to be in units of 12 cubits (about 6 m). This should imply that the whole plan was drawn to scale, but it is difficult to establish from archaeological data whether this was so. Another plan from Nippur, of the same date, shows what are clearly fields, canals and a river bend.[4] This type arises mostly from property sales or disputes, for which a visual record would be simpler than an enumeration of boundaries. Because of the durability of clay tablets, more of these have survived than of comparable documents from Classical antiquity.

Maps covering a wider area of Mesopotamia seem to be confined to two tablets, one of which from Nippur has nine towns or villages linked by a road and canals.[5] In this case the layout of artificial waterways may have been the prime concern. But the best example is a tablet found at Yorghan Tepe near Kirkuk, datable to *c.* 2300 BC.[6] It gives a map of the area round Gazur (later Nuzi), with a plot of 127 ha. (314 acres) in the middle, together with owner's name, but also hills on each side and a river or canal between. Perhaps the most significant feature is the inscribed orientation, of which this is the earliest extant example, with the words for 'west' at the bottom, 'east' at the top and 'north' to the left.

The Babylonian World Map ('map of the four areas of the world' seems to be its title), which is in the British Museum, is, although small, very famous.[7] Its date is very late in Mesopotamian civilization, about 600 BC, and although some scholars have claimed it as copied from a much earlier map, this cannot be demonstrated. Its identity as a world map is proved by the adjacent text, which mentions seven outer regions beyond the encircling ocean. This is a slightly different concept from that of the early Greeks, for whom the encircling ocean was outside all known lands. Although the interpretation is uncertain, it is thought that one phrase refers to a region where the sun is invisible. If this is correct, it might imply exploration up to very northerly latitudes; such exploration could have been easier at the date of the map than earlier. The seven distant regions were marked with triangles (not all preserved), attached to which are figures thought to refer to distances between them, though nothing definite can be inferred. West seems to be at the top: Babylon is marked in the centre, with lines leading to mountains and the swamp of lower Mesopotamia. Assyria is to the right of Babylon, with what may be Urartu (eastern Turkey and Armenia) above Assyria. Attempts have been made, not very successfully, to identify other place-names. But great accuracy is not to be expected in such a map, whose main purpose was to give an idea of the location of the outer areas of the world, visited by legendary heroes.[8]

The Babylonians were noted mathematicians and astronomers. In the former field, among other things, they attained a remarkably close approximation for $\sqrt{2}$, namely 1.414213.[9] Our divisions into 60 and 360 for minutes, seconds and degrees are a direct inheritance from the Babylonians, who thought in these terms. They had a sexagesimal notation, e.g. 70 was expressed as 1,10. Babylonian lore was passed down to the Greeks by Berosus (*c.* 290 BC) and others.

In so far as astronomy is concerned, Babylonian researches were normally recorded in writing, not in cartographic form. The nearest approach lies in circular tablets which plot the positions of groups of stars. Barely verging on the cartographic are diagrams which divide a circle into twelve segments, corresponding to the twelve months, each containing an indication of the stars visible from Babylon in the appropriate month. Some of these diagrams were not the result of scientific placing of observations but connected with horoscopes or magic.

PALESTINE

Although we have no Phoenician maps, the influence of Phoenician sailors on Greek cartography should not be ignored.[10] In particular they applied detailed astronomical knowledge to navigation. It is uncertain whether maps were used in ancient Palestine:[11] if they were, they must have been on papyrus and have not survived. A number of Old Testament texts give boundaries of tribes, and in Genesis 10 we find an enumeration of the tribes of the known world based partly on relationship and partly on geographical location.

EGYPT

Egypt was undoubtedly a land of accurate measurement.[12] From earliest times much of the area covered by the annual Nile floods had on their retreat to be re-surveyed in order to establish the exact boundaries of properties. The survey was carried out, mostly in squares, by professional surveyors with knotted ropes. However, the measurement of circular and triangular plots was envisaged: advice on this, and plans, are given in the Rhind Mathematical Papyrus of *c.* 1600 BC (British Museum).[13] The Great Pyramid, on a square ground-plan, was built not only with precise orientation to the four compass points but with very little difference in the dimensions of the sides.

In so far as cartography is concerned, perhaps the greatest extant Egyptian achievement is represented by the Turin papyrus,[14] collected by Bernardino Drovetti before 1824. This is a painted map of *c.* 1300 BC, now in a number of fragments, which shows hills between the Nile and the Red Sea containing gold and silver mines and the roads in the area. An inscription on the map, in hieratic, the cursive form of hieroglyphic, reads: 'The hills from which the gold is brought are drawn red on the plan'. Other inscriptions include: 'peak' (on the red area), 'cistern', 'abode of Amun of the pure mountain', 'houses of the gold-mining settlement', 'road leads to the sea'. A prominent feature of the plan is what seems to be a winding wadi, of about the same width as the roads. The map was drawn in connection with a statue of a pharaoh which had never been completed. The gold mines have been located at Umm Fawakhir in the Wadi Hammamat. But it is misleading to think of the Turin papyrus as the first geological map. It was not scientifically compiled, and although the use of colour shows originality, its application seems to have been legal.

Nevertheless, we do not possess anything similar from antiquity; indeed from the Graeco-Roman world the only plan of a mine, and that only probable rather than certain, is the incised rock at Thorikos, Attica, mentioned below (p. 26). There is, however, from Egypt, about fourteen hundred years after the Turin papyrus, a theoretical plan of tunnelling. This is in manuscripts of the *Dioptra* of Heron of Alexandria,[15] whose object was to demonstrate the use of geometrical theorems in tunnel surveying. This could have served either for aqueducts or for mine shafts. An early Greek accomplishment in the sphere of aqueduct

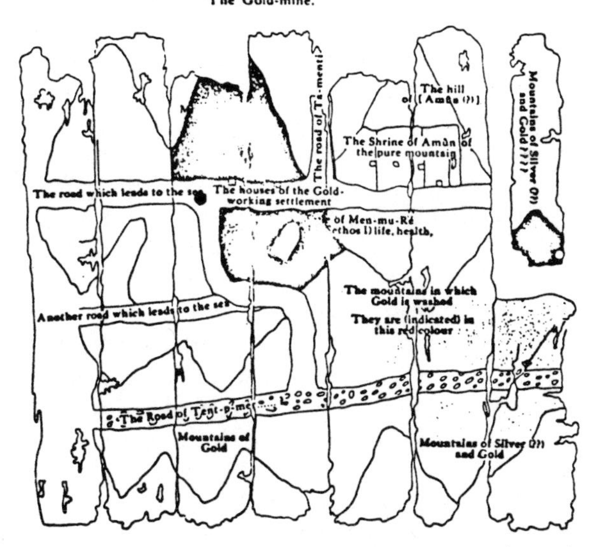

Fig. 1. The Turin Gold Mine Papyrus: fragmentary coloured map in the Museo Egizio, Turin, showing an area of gold-mines between the Nile and the Red Sea. Hieroglyphic captions translated into English: J. Ball, *Egypt in the Classical Geographers* (Cairo, 1942).

engineering is the long tunnel of Eupalinus at Samos commissioned by Polycrates (tyrant *c.* 540–*c.* 522 BC),[16] so that in this field, apart from the resources of the Alexandrian Library, Heron had Greek as well as Roman work on which to base his theory. But it is tempting to think that in mining survey he may have learnt something from dynastic Egypt.

Other topographical maps are mostly drawn in what seems to us a combination of the primitive and the formal, partly in plan and partly in profile. The commonest subject is gardens,[17] which show paths surrounded by date palms and sycamores; there are also pools with ducks and water plants, and outside the garden areas orchards or vineyards. Persons and animals are represented in elevation as on an Egyptian picture, while trees, although also represented in elevation, sometimes appear upright, sometimes, if on a border, turned through an angle of 90°. This combination of plan and elevation may be seen in plans of buildings dating from 1500–1000 BC (one from Deir el Bahri shows a shrine and precinct wall, with measurements), in which the ground-plan of a house may be combined with elevations of porticoes and items in storerooms.[18] Such a combination is also visible in a painting from Egyptian Thebes showing an army drawn up for battle and a town surrounded by moats.[19] A Mesopotamian parallel is a Nineveh bas-relief of the sixth century BC, which shows the city of Madaktu and round it a triumphal procession celebrating its capture by Assurbanipal.

Religious geography results in somewhat similar maps. In the *Book of the Dead* (*c.* 1400 BC) there are representations of the ideal plot of land which the deceased will till in the kingdom of Osiris.[20] This is rectangular and intersected by canals, and the use of colour makes it more vivid. Painted coffins of *c.* 2000 BC from El-Bersha illustrate routes to reach the gateway to the Underworld.[21] The upper route, in blue, is by water, the lower one, in black, by land. This collection of illustrated spells has in modern times been called *The Book of the Two Ways*. Linked with this is the day and night journey of Re, the sun god.

Cosmographical illustrations from Egypt with map-like features are not uncommon.[22] In royal tombs of the eighteenth to twentieth dynasties the goddess Nut is shown bending over to portray the universe; under her is Shu, who occupies the space between earth and sky, and Geb, the earth god. The feet of Nut touch the east horizon, while her arms hang down to the area of the setting sun. We also find the sun passing through Nut's body, with the names of foreign lands inscribed on the edge of the diagram. This commemorates the legend that Nut swallowed the sun every evening and gave it rebirth every morning.

Celestial cartography is represented by astronomical ceilings from the fifteenth century BC onwards, giving decans (literally 10° of a zodiacal sign, hence timings of appearance and disappearance of stars), constellations and planets.[23] In some of these a diagram shows stars for the purpose of calculating night hours. But this type of ceiling, which at times illustrates a hippopotamus or a crocodile, is not to be thought of as very scientific. In Egyptian celestial mapping the twelve signs of the zodiac appear only late and evidently as an introduction from Babylonia or elsewhere.

It is evident that the Egyptians were very familiar with large-scale cartography and that the makers of these plans treated their subject in a neat and formal way typical of dynastic art. One may hope that more papyri containing such maps and plans may be found in future excavations. Despite the prevalence of re-survey, no survey maps have survived from dynastic Egypt. From Ptolemaic Egypt there is a rough rectangular plan of surveyed land accompanying the text of Lille Papyrus 1, now in Paris;[24] also two from the estate of Apollonius, minister of Ptolemy II: one has a plot, the other a palisade; both contain canals, and each has some means of orientation.[25] Sophisticated survey techniques were introduced by Alexandrian applied mathematicians, but to what extent they were used in practice we do not know.

EARLY QUASI-MAPS

The links of dynastic Egypt with Minoan civilization are well established. Although we have nothing approaching a map or plan from Minoan Crete, there is in the National Museum, Athens, a map-like fresco from the island of Thera (Santorin), which until its destruction through the catastrophic eruption of its volcano about 1500 BC was part of the same well developed civilization.[26] In the so-called 'House of the Admiral' at Akrotiri, Santorin, was discovered this excellently preserved wall-painting of *c.* 1600 BC. It represents, in oblique perspective, views of a seaside community and the adjacent sea. One cannot exactly speak of it as a map, but there are several map-like elements: a winding river, with palm trees and other vegetation, and a mountain, the areas round both of these also containing animals; a coastal outline, including a small settlement; and the harbour of this settlement, with much shipping out at sea. The fact that many of these landscapes also have groups of human beings, performing different activities in different places, tends to detract from the map-like quality. There has been discussion (a) whether the ships are engaged in battle or not, (b) whether the location is on the coast of Crete or of north Africa.

Whereas the wall-painting described above is thought to have been the product of a literate society, since contemporary Crete, Mycenae and Pylos were literate, we have to turn to non-literate societies for anything similar of early date further west. The most elaborate are the Bronze Age rock incisions in the Val Camonica, northern Italy (p. 102). Whereas the Iron Age incisions stress the pictorial side, these earlier ones are more in the form of plans. Many smaller rock carvings, some of which can be thought to be plans, were observed in the Ligurian Alps in the late nineteenth century by the Rev. C. Bicknell, founder of the Bicknell Museum at Bordighera.[27] Plan-like elements may also be detected in the stele of Novilara, of about the seventh century BC, in the Museo Oliveriano, Pesaro.[28] Although in this there seems no doubt that the largest incisions portray a sailing ship and two smaller vessels, the interpretation of the rest, e.g. river or snake, and how much may be thought to be in plan, are very debatable.

One might have supposed that the Etruscans, a well organized, artistic and religious seafaring people, would have left maps or plans. They were indeed very conscious of orientation, and had at Marzabotto a street grid system in long, narrow rectangles similar to those of Greek colonies in southern Italy and Sicily. On the details of orientation, Roman writers who did not know Etruscan (a non-Indo-European language, written from right to left[29] in an adaptation of Greek lettering) are not very reliable. The elder Pliny says that the Etruscans divided the heavens into sixteen (4 × 4) parts, calling the eight eastern subdivisions 'left' and the eight western 'right'.[30] This implies that the soothsayer would face south in his observation of the heavens. But usage may have varied between celestial and terrestrial observations. Frontinus writes: 'The origin of centuriation, as Varro mentions, is in the Etruscan lore [*disciplina Etrusca*], because their soothsayers divided the earth into two parts, calling that to the north "right" and that to the south "left".'[31] This implies, as one might expect in such a ritualistic society, that Etruscan surveying was carried out by, or under the guidance of, soothsayers and that they faced west.

The only extant rough plan of Etruscan origin, on the vase from Tragliatella (p. 147 below) showing a maze with a 'game of Troy', throws no light on this problem. But a religious object of about the third century BC, the Piacenza bronze liver, certainly has some bearing on orientation, though perhaps not on mapping.[32] It is a bronze model of a sheep's liver, 12.6 cm long, found in 1877 between Settima and Gossolengo, and now in the Museo Civico, Piacenza. On its top it has projections based on those of a sheep's liver, together with a flat section divided into boxes representing zones, each labelled with the name of an Etruscan deity. The convex base has two sections, labelled with the Etruscan words for 'sun' and 'moon'. The connection with soothsaying from the inspection of livers is obvious, and a parallel may be drawn with a Chaldaean terracotta liver in the British Museum.[33]

Since the section round the part of the liver representing the *processus pyramidalis* is roughly rectangular, some scholars have thought that a rectangular scheme of land division is intended to be shown. This is very uncertain, although the orientation might suit land better than sky as far as Etruscan practice went. An alternative theory is based on an observation by Martianus Capella, writing in the early fifth century AD.[34] After saying that the ancients divided the heavens into favourable and unfavourable areas, he gives Roman names of deities associated with such areas. Some of these

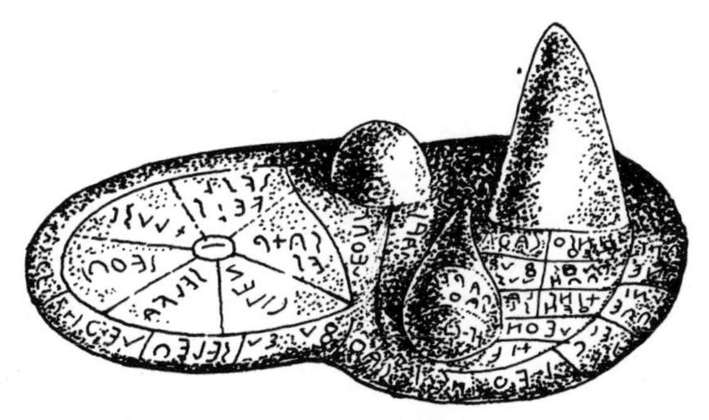

Fig. 2. The Piacenza bronze liver: Etruscan model of a sheep's liver, 3rd century BC, used for divination. To right, the *processus pyramidalis*; round the rim, names of deities, e.g. VNI (right to left) = Juno.

correspond with the names of Etruscan deities on the liver; but no general correspondence can be worked out, and the orientation is less satisfactory. The Piacenza liver, one may infer, is not a map: at best it may be described as a schematic model.

In the Graeco-Roman areas no world maps of the early period, or approximations to them, have been discovered. If they, or even *periploi* (p. 130), had existed then, poets might not have recounted the sea voyages of such heroes as Jason, Odysseus or Aeneas as so circuitous. But something akin to a map is described by Homer in the *Iliad*. It is a well-known feature of Homeric epic that some descriptions mirror in poetic language Mycenaean institutions of many centuries earlier. This may be the case with what seems to us a somewhat unusual description, that of Achilles' shield.[35] It was made for that hero, the poet tells us, by the craftsman god Hephaestus, of golden, tin and bronze plates. On one of the bronze plates he engraved countless subjects:

> Earth, sky and sea, the never-tiring sun,
> The moon when full, all stars that deck the sky:
> Orion's might, Hyades, Pleiades,
> The Bear, known also to us as the Wain,
> Which, circling round, watches Orion close,
> Alone not sinking into Ocean's stream.[36]

After this he is said to have engraved two fine cities, one of peace and one of war. Agricultural scenes, dancing and merry-making are among the aspects of everyday life depicted. Round the whole is the great strength of the river Ocean.[37] We cannot know whether this description reflects actual usage of any kind. The only shield-map known to us, the Dura Europos shield (p. 120), dates from some one thousand years later and served a practical purpose. But the idea of one celestial scene, with the constellations depicted, perhaps to enable one to find one's way at sea, and one terrestrial, with cities surrounded by the Ocean, does suggest early attempts at map-making which may have been partly misunderstood in the Dark Ages that followed the Mycenaean period. Homer is said to have lived either on the west coast of Asia Minor or on an adjacent island. Since Miletus was the birthplace of Greek map-making, and since Homer not only showed a keen awareness of geography but evidently lived not far from there, it is appropriate that many later Greeks should have thought of him as the father of geography.

CHAPTER II

EVIDENCE FROM ANCIENT GREECE

It is unfortunate that so little remains from what was clearly a great early period of the development of mapping. This may be due to the perishable or reusable materials on which maps were drawn. What evidence there is suggests that these were normally either painted on wood or more rarely engraved on bronze. Manuscripts of Greek authors of the Classical and Hellenistic periods, copied mostly from the ninth to the fifteenth century, survive either without maps or, in the case of Aristotle's *Meteorologica*, with medieval maps which may or may not reflect classical sources.

It is from the early philosophers that Greek mapping concepts spring. Although Greek colonization has left conspicuous traces of urban and rural land division, there is no evidence that surveyors of these colonies used maps. Moreover the part played by Greek navigators in map-making is disputable: Greek *periploi* appear to have been verbal instructions. The Ionian philosophers and their successors were interested in theoretical rather than practical cartography; though it will be seen that their researches could have practical effects. Their study of cosmology led a number of them to map the heavens as much as the earth. Although this book is in principle concerned with terrestrial maps, the achievements of the Greeks cannot be completely understood without some recognition of the work of their philosophers and mathematicians on the mapping of the heavens.

CELESTIAL MAPPING

Evidence of early interest in celestial cartography is scanty. When we are told by Diogenes Laertius that Anaximander (see below), the first acknowledged cartographer, was also the first to construct a sphere,[1] we may imagine that this was a sphere of the heavens.

Something similar may have been designed by Parmenides or the Pythagoreans in southern Italy. But the real step forward came with Eudoxus of Cnidos (*c.* 408–355 BC), the well-known astronomer, who was a pupil of Plato's.[2] His great achievement was to construct a globe showing the sky as seen from outside. He explained this globe in his works *Phaenomena* and *The Mirror* (*enoptron*). Its impact on Greek and Roman readership was assured much later when Aratus of Soli (*c.* 315–240 BC) wrote a verse rendering of the *Phaenomena*.[3] From this we can see that the chief feature was the location on the globe of conventional signs for the heavenly bodies depicted. Aratus' version was so popular that it was several times translated into Latin verse. No fewer than three of these renderings are extant, by Cicero, Tiberius' nephew and adopted son Germanicus, and Avienius (p. 141). A very generalized idea of the appearance of the globe may be obtained from the Farnese Atlas in the Naples Archaeological Museum.[4] The supporting of the sky by the giant Atlas was a favourite theme of Greek art, and this is a Roman copy of a Hellenistic original.

The last scientist in a politically independent Greek state, Archimedes of Syracuse (287–212 BC) was famous as inventor of the theory of displacement, but he did contribute much to celestial mapping.[5] He made many spheres; one of these seems to have been a solid globe which showed the people, animals etc. after which all constellations were named.[6] It was very accurately designed and in bright colours. Another was a machine akin to an orrery, imitating the motions of the sun, the moon and five planets. It was specially contrived to show solar and lunar eclipses after the correct number of revolutions. Archimedes was unfortunately killed by a Roman soldier at the siege of Syracuse as he was making geometrical drawings in the sand. But he and his drawings were a major military target, as it was he who had devised cunning war-machines used against the Romans. The 'orrery' went as spoils to the commander-in-chief, M. Claudius Marcellus. Archimedes' work was obviously appreciated, as his globe was placed on public view in the Temple of Virtue in Rome.

TERRESTRIAL MAPPING: CLASSICAL

Later Greeks considered that the first map-maker was Anaximander of Miletus (*c.* 611–546 BC), whose master Thales was said to have visited Egypt to consult priests, and to have predicted an eclipse of the sun. Miletus, as a Greek city[7] in Asia Minor, was well placed to absorb aspects of Babylonian science, including possibly the gnomon, the upright member of a sundial, though Anaximander is said himself to have invented it. He is also said to have set up a sundial in or near Sparta.

Anaximander was the second philosopher of the Ionian school, which was particularly interested in cosmology. The shape and size of his map are difficult to envisage from the accounts.[8] It was evidently a map of the known world, and Strabo and Agathemerus call it a *pinax*, a term used particularly of painted panels but sometimes also of bronze panels. Diogenes Laertius says that it portrayed an outline (*perimetron*) of the land and sea.[9] But when the same author tells us that Anaximander believed in a geocentric universe with a spherical earth, he is suspect, since the spherical concept was not devised early, and others attributed to Anaximander the different concept of a cylindrical earth.[10] Nevertheless, he may indeed, as Diogenes Laertius claims, have been the first to construct a sphere, though more likely celestial than terrestrial.

The practical map-making which developed from Anaximander's map may be illustrated from a well-known story in Herodotus.[11] In 499–8 BC, Aristagoras, tyrant of Miletus, made a tour of important cities on mainland Greece looking for allies against Darius I, King of Persia. He took with him on this tour what Herodotus calls 'a bronze tablet [*pinax*] with an engraving of a map [*periodos*, literally "going round"] of the whole world with all its rivers and seas.' Among his contacts was King Cleomenes of Sparta, and on it he showed him all the areas on the way from Ionia to Susa, capital of Persia. The last region of Asia Minor on the proposed march, Cilicia, is described as 'opposite Cyprus', implying that Cyprus too appeared; and the regions east of Asia Minor are given as Armenia, Matiena, and Cissia with the city of Susa. Like many other maps in antiquity, however, this presumably had no scale; for Cleomenes, two days later, asked 'How long would such a march take?' 'Three months', was the reply, whereupon despite attractive offers of money he refused. This map was probably developed from that of Anaximander. But we may presume that it also contained the course of the Royal Road, which Herodotus describes in some detail immediately after, giving the number of staging-posts and the distances. This road had been carefully measured for the Great King by road surveyors; and the general proportions of Aristagoras' map, particularly the section relating to Asia, may well have been guided by such survey work on it. A plausible theory is that the geographer and mythographer Hecataeus of Miletus (*fl.* 500 BC) was the indirect promoter of this map, based on Anaximander and on his own travels in Asia and Egypt. His *Periodos Gēs*, 'Journey round the world', dealt with Europe in Book I, Asia (in which he included Africa) in Book II.[12] Some three hundred fragments exist, but they are mostly brief and not very informative as recorded by Stephanus of Byzantium. When we are told that he enormously improved Aristagoras' map, this probably means that he criticized it in his text rather than re-drew the map.[13]

To judge from the term *periodos*, 'way round', Aristagoras' world map is likely to have been circular. In another passage (iv. 36), Herodotus remarks: 'I am amused to see so many people producing circular maps for no good reason. We are shown the Ocean flowing round the earth perfectly circular as if turned on a lathe, with Europe and Asia the same size.' He, like some other Greeks of the Classical period, thought that Europe was much larger. The idea of an encircling Ocean was a very old one, perhaps inherited from early Babylonian maps and reinforced by Greek mythology as interpreted by Homer. These early Greek maps had Greece in the centre, perhaps with Delphi occupying the central position. This was not only because it is fairly centrally situated in mainland Greece, even claiming to be the *omphalos*, 'navel', of the earth: it was also the chief religious meeting-place of the Greeks, where Apollo's oracle advised on such matters as colonization.[14] Herodotus' criticism enables us to see that map production by the time he was writing (*c.* 444–430 BC) was fairly large, even though repetitive.

A few chapters later (iv. 42) Herodotus expands on his view of how maps should be modified. Before surrounding Europe and Asia with the Ocean, he says, we should examine what is known about them. In the case of Africa, which he calls Libya, Pharaoh Neco (609–594 BC) had sent Phoenicians who in several ships circumnavigated it in three years, thus proving that except at the Isthmus of Suez it was surrounded by water. When, after their clockwise circumnavigation, they returned to Egypt via the Pillars of Hercules (Straits of Gibraltar), they reported that they had the midday sun on their right. Despite this obvious evidence of sailing south of the equator they seem to have been disbelieved. Herodotus' views of the extent of the three continents are discussed on pp. 57–8 below.

But already before Herodotus the idea of a spherical earth had led to a new cartographic concept. He himself emigrated to Thurii in southern Italy *c*. 443 BC. This was not far from Croton, to which about 530 BC Pythagoras had moved from Samos to found a mathematical and philosophical community. Although Pythagoras wrote nothing, we know that he considered the sphere as the perfect shape for all bodies in the universe, presumably including the earth. This was taken up by Parmenides of Elea in southern Italy (born *c*. 515 BC), said to have been the first[15] to divide a spherical earth into five zones, one hot, two temperate and two cold. It seems likely that he illustrated his division either on a map or a globe.

If, however, a flat surface was used to portray the inhabited world (oikumene), the philosopher map-makers debated which was the best shape for such a map. An oblong or oval shape was suggested by Democritus of Abdera (*c*. 470 or 460–370 or 360 BC), who with his master Leucippus also drew up the concept of the atom. Democritus travelled very widely, consulting learned men in Egypt, Mesopotamia and even, according to one source, India. His conclusion on the shape of the inhabited world was that the proportion of its length (east–west) to its width was 3:2.[16] This proportion had some influence on subsequent cartographers of the oikumene. But it was recognized as being based on incomplete knowledge; Plato makes Socrates say that outside the world known to the Greeks there are probably a great number of people living in a great many similar regions.[17]

At this point we may consider what knowledge or use of maps the man in the street had in Classical Greek times, for example at the time of the Sicilian expedition, 415 BC. Did the ordinary Athenian know where Sicily was, and had he any idea of the details of its topography? Evidently the answer to the first question is yes, to the second no. On the one hand Plutarch tells us that just before the invasion the average Athenian could sketch the outlines of Sicily and place it in relation to north Africa in general and to Carthage in particular.[18] On the other hand we are told by

Thucydides that the man in the street generally knew little about the size or population of Sicily.[19] Evidence from Aristophanes is less satisfactory, but after all it was part of the comedian's stock in trade to exaggerate and caricature. To start with, in *Clouds* 200ff., written in 423 BC, Strepsiades the simple countryman, when he is shown a map of the world, imagines it is a plan of an allotment. Then, when he realizes what it is, he has difficulty in recognizing Athens, Euboea and Sparta; but the reasons are expressed satirically. Where Athens appears on the map, he cannot see any of the hordes of jurymen. The island of Euboea is surprisingly long on the map, but even so he reflects that it is being outpaced, i.e. politically outdistanced, by Athens. Finally, enemy Sparta looks much too close for any Athenian's peace of mind.

The allusion by Aristophanes to an allotment plan is revealing. That is the sort of plan that many countrymen might be expected to have seen. Plans are liable to be produced, among other things, for amplification of legal definitions. Such plans might well have been on papyrus; from ancient Greece, as opposed to Egypt, Herculaneum and elsewhere, no papyri have survived. The material most likely to be preserved is stone; but whether one such from the Athens area is more than a short inscription is uncertain.[20] It might also contain an extremely small plan or a monogram; its heading is 'boundary between shop and house'. At Thorikos, east Attica, just above the west parodos of the theatre, is the entrance to a fourth-century BC mine recently explored by Professor H. F. Mussche and others of the Comité des Fouilles Belges en Grèce. On the rock face immediately above this entrance is what appears to be a small plan of the mine, corresponding roughly to the 120 m section so far excavated. Such an example is very easy to miss, and there may be others similar which are as yet undetected. The interest of Thorikos in surveying is indicated by three inscriptions which read ὅρος οἰκοπέδων, 'boundary of apportionment', cut out on the rock.[21]

The fourth century BC was one of great scientific achievement. Although Eudoxus' contribution to terrestrial mapping was probably smaller than his celestial contribution (p. 22), we are told by Strabo that he was regarded as an expert in figures and 'climates', i.e. latitudes, and by Agathemerus that he regarded the length of the inhabited world as double its breadth.[22] 'Figures' (*schemata*) must refer to geometrical figures relative to terrestrial cartography. 'Climates' (*klimata*,[23] literally 'inclinations') are not to be understood in the modern sense, but as latitudes or latitudinal zones often based on maximum hours of sunshine. This use of *klimata* descended via Ptolemy's criticism of Marinus, with modifications, to the Middle Ages and the Renaissance. Clearly the proportion of the inhabited world mentioned above was used for a long time in antiquity. In the first century BC, Geminus of Rhodes wrote: 'The breadth of the inhabited world is approximately half its length; so to draw a map to scale one should use a rectangular panel, with length twice its breadth'.[24] Whether Eudoxus had such a scheme of rectangular panels we do not know: the surviving fragments concern physical geography or are descriptive of peoples of the world. But the idea of a rectangular basis may also underlie the approach of his contemporary, the historian Ephorus.

Unfortunately our evidence for this is extremely late, coming as it does from Cosmas Indicopleustes (p. 171). In Book IV of his History, Ephorus of Cyme in the Aeolid (*c.* 405–330 BC) equated compass points, expressed in terms of winds, with peoples, thus:

Wind	*Direction*	*People*	*Area occupied*
Apeliotes	East	Indians	From summer rising to winter rising
Notos	South	Ethiopians	From winter rising to winter setting
Zephyros	West	Celts	From winter setting to summer setting
Boreas	North	Scythians	From summer setting to summer rising

The areas allocated to Ethiopians and Scythians indicated that they occupied a greater arc of the circle than Indians and Celts, though the whole scheme is only very approximate. Cosmas Indicopleustes says that Ephorus illustrated this concept 'with the help of the following drawing', whereupon a rectangular map follows, with Greece or the Aegean in the centre.[25] Although some of his details are either Christian in concept or too crude for Ephorus to have incorporated, we may at least conjecture that Ephorus, like Eudoxus, favoured a rectangular layout for a map of the known world.

Whereas a number of previous writers had assumed the earth to be spherical, Aristotle (384–322 BC) set out proofs of sphericity: at lunar eclipses the earth's shadow on the moon is circular; the celestial pole rises as one travels north. He adopted Parmenides' five zones, but defined them in terms of equator, tropics and arctic circles. Like Herodotus, he criticized circular maps of the oikumene, and he gave the proportions of the latter as over 5:3 (Straits of Gibraltar–India, Ethiopia–Sea of Azov). He was pessimistic about further exploration either north–south, because of adverse climatic conditions, or east–west, because of the enormous stretch of ocean between India and the Pillars of Hercules (Straits of Gibraltar). Aristotle's estimate of the circumference of the earth was 400,000 stades. We do not know what length of stade he used, but the extremes are approximately 60,000 and 80,000 km (*c.* 37,000–50,000 miles), so in any case his estimate was appreciably larger than Eratosthenes' measurement (p. 32).

The only passage in Aristotle where we know that there was a diagram akin to a map either drawn or intended is *Meteorologica* ii.6. Modern editions[26] reconstruct this, and it also exists in corrupt form in a Madrid manuscript of the twelfth century.[27] The object is to show the positions of the winds, so the centre of the map is where we live, in his case Greece or the Aegean, and the circle represents the horizon as seen from that point. By marking on the circumference, in addition to the cardinal points, the summer and

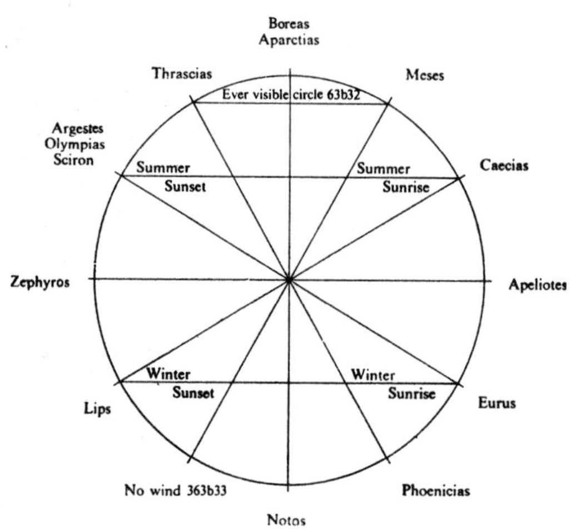

Fig. 3. Reconstruction of wind map, Aristotle, *Meteorologica* ii.7. Eleven points are indicated from which named winds come.

winter risings and settings of the sun, and joining these points, he was able in effect to depict the north temperate zone; his diagram must have been purposely incomplete, as he recognizes only ten or eleven winds, not twelve. As heirs to the tradition of this Aristotelian wind map, we may compare the Pesaro anemoscope (p. 110) and the map attached to Ptolemy's *Handy Tables* (p. 170). The same Madrid manuscript which gives the diagram from Aristotle's *Meteorologica* also has attempts at showing in map form where certain prominent mountains and rivers are situated. As these contain serious topographical errors, they are unlikely to be derived from Aristotle's diagrams: rather they must be poor reconstructions from his text of the *Meteorologica*.

This was the state of world cartography when Alexander the Great, himself a pupil of Aristotle, set out on his expedition to the East which resulted in the overthrow of the Persian Empire (334–323 BC). Although he believed himself to have a divine mission, military conquest was not by any means his sole aim. He had a genuinely enquiring mind, and wanted if possible to explore the whole land-mass to the East. He instructed secretaries to collect whatever material was available on the more inaccessible regions, and took various scholars with him to write up the areas covered. Whether the collected material included maps we are not told. But the technicians on the expedition included two road-measurers[28] whose function was to keep a record of all distances between stopping-places (the elder Pliny gives some details of these) and to describe the geography, soil, flora and fauna of all areas traversed. This information was incorporated in a daily expedition report (not extant) compiled by Eumenes of Cardia. It is clear that later topographical writers, such as Isidore of Charax (p. 124), drew extensively on the itinerary content of the expedition. Alexander's plan to go further and explore the Far East was thwarted by a revolt of his army. Rather than abandon all discovery of new territories, he substituted a land and sea journey from the Indus delta to the head of the Persian Gulf (pp. 134–5).

In the West, about 330 BC, Pytheas, a Greek from the colony of Massilia (Marseilles), explored the Atlantic coasts of Europe as far north as possible.[29] Although in antiquity geographers and others were sceptical of his findings,[30] we can see that his voyage was valuable in scientific terms. His main object seems to have been to work out latitudes for many of the remoter places, either through the length of the longest day or through the height of the sun at the winter solstice. He knew the exact location of the celestial pole, and worked out the latitude of Marseilles as 43°12′ N., very near the correct value of 43°15′.

His voyage (p. 136) took him via Cadiz, up the Spanish coast, to the Cassiterides or Tin Islands, whose location is disputed. He then circumnavigated the British Isles, and may even have sailed into the Baltic. At the furthest point north at which his observations are unquestionably recorded, Mona (here the Isle of Man[31]) and the Bight of Lübeck, the sun rose to only 6 cubits = 12° at midwinter, and the longest day had nineteen equinoctial hours. But from the geometry of the sphere he could tell that at some point the sun must be constantly visible at midsummer, giving the longest day twenty-four hours; so at that point he placed (on a map or in his text) an island that he called Thule. Since his own account has not survived, we cannot tell whether he visited such an island or whether it can indeed be related to an actual piece of land.

Dicaearchus of Messana (Messina), *fl. c.* 320 BC, a pupil of Aristotle, wrote among other lost works *Periodos gēs*, which can be taken as meaning either 'voyage round the world' or 'mapping of the world'. He started from Democritus' dimension of 3:2, and worked on a base line (*diaphragma*) from the Pillars of Hercules via Sardinia, Sicily, where he was born, the Peloponnese, where he then lived, southern Asia Minor, the Taurus mountains and the Himalayas. This base line divided the width of the known world very approximately into two equal halves. On it or branching out from it he gave estimates of lengths, thus:

From	*To*	*Stades*
Pillars of Hercules	Straits of Messina	7000
Straits of Messina	Peloponnese	3000
Peloponnese	Head of Adriatic	over 10,000

Although Strabo (64/63 BC–AD 21 or after) criticized Dicaearchus' figures,[32] the intervening three hundred years, with the resources of Rome to hand, had supplied much more careful measurements.

How far these writers made or looked at maps cannot be determined, but they were accumulating information on the framework as well as some details of lands of the known world. A sidelight on the display of maps in Athens comes from the will of

Theophrastus (*c.* 370–*c.* 286 BC), successor to Aristotle as head of the Peripatetic School.[33] Evidently with the general aim of educating the public of Athens, he requests that 'the panels [*pinakes*] showing maps of the world [*periodoi gēs*] should be set up in the lower cloister'. These were evidently wooden panels, which like pictures could easily be removed when required. This information illustrates the wide spread of world maps. Coins struck about this time on Rhodes (p. 146) and showing a small part of Asia Minor are simple forms of regional maps.

TERRESTRIAL MAPPING: HELLENISTIC

An important by-product of the capture of Egypt by Alexander the Great, and the establishment of his general Ptolemy I Soter, was the planning of Alexandria, among other things as a new centre of Greek learning. During the reign of the successor, Ptolemy II Philadelphus (308–246 BC, on the throne from 283/2 BC), the Alexandrian Library and Museum were planned. The library was particularly valuable as having a very good collection of up-to-date scientific works.[34] Apollonius of Rhodes wrote there and in Rhodes his *Argonautica*, a largely geographical epic poem. In one passage, evidently as a result of researches he had made, he claims that in early times a party of emigrants from Egypt to Colchis, on the east coast of the Black Sea, erected pillars on which some sort of map of their land and sea journey was etched.[35]

Timosthenes of Rhodes, one of Ptolemy II's admirals (*fl.* 270 BC), wrote a treatise, now lost, *On Harbours*.[36] He added two winds to the ten given in Aristotle's *Meteorologica*, and allocated remote peoples or countries to these twelve directions. Since we know that he placed Scythia beyond Thrace and Ethiopia beyond Egypt, it seems likely that he made Rhodes the centre of his windrose, which may have been accompanied by a map; this was adopted by many of his successors, and even after the fall of the Roman Empire we may compare what may be deduced from the Ravenna Cosmography (p. 174). Rhodes was a good place from which to make observations, as it had a long naval tradition and must have built up a mass of nautical information gathered over a long period. Nevertheless, Strabo was easily able to find flaws in Timosthenes' geographical descriptions; thus he placed Metagonion (Melilla, north Africa) opposite, i.e. due south of,

Marseilles, whereas it should, says Strabo, be placed opposite Nova Carthago (Cartagena).

A vital contribution to reality in mapping on a world scale came with a scientific estimate of the circumference of the earth. Its originator accepted that the earth was a sphere, and assumed that that sphere was perfect. Eratosthenes (*c.* 275–194 BC), who was born at Cyrene and studied in Athens, was invited by Ptolemy III Euergetes, King of Egypt 246–221 BC, to come to Alexandria as tutor to his son and, shortly after, director of the Library. His relevant works, neither of which has survived, were *On the Measurement of the Earth* and *Geographica*:[37] Cleomedes summarizes the former and Strabo criticizes the latter. While still keeping to the geocentric views of the universe, he started from the assumption that the sun was so distant that for practical purposes one could consider its rays parallel anywhere on earth; and it was this theory that enabled him to arrive at a remarkably accurate calculation. He assumed that Syene (Aswan), where at midday on the summer solstice the sun was exactly overhead, was on the same longitude as Alexandria, though there is a difference of 2°. He worked out the angle, at Alexandria on the summer solstice, between the vertical and the angle of the sun at midday as 1/50 of 360°. Then the angle subtended at the centre of the earth by Alexandria and Syene would be equal to this angle. If these places were approximately 5000 stades apart, the circumference of the earth would be 50 × 5000 = 250,000 stades. But as a mathematical ploy, in order to achieve a number divisible by 60 or 360, so as to correlate stades with his subdivisions or degrees, he emended this to 252,000 stades.

A stade (*stadion*), originally the distance covered by a plough before turning, was 600 feet of whatever standard was used.[38] Scholars have disputed what length of stade was used by Eratosthenes. A late writer, Julian of Ascalon, says that Eratosthenes and Strabo both had 8⅓ stades to a mile. This is not true of Strabo, who had 8 stades to a Roman mile.[39] Eratosthenes never reckoned in miles, but we may presume Julian was defining an Olympic stade of 178 m 60, which would make the circumference of the earth 45,007 km (27,967 imperial miles) as against the actual equatorial circumference of 40,075 km (24,902 miles). An alternative suggestion, made in the nineteenth century, which is based on Egyptian measurements and is thought to have

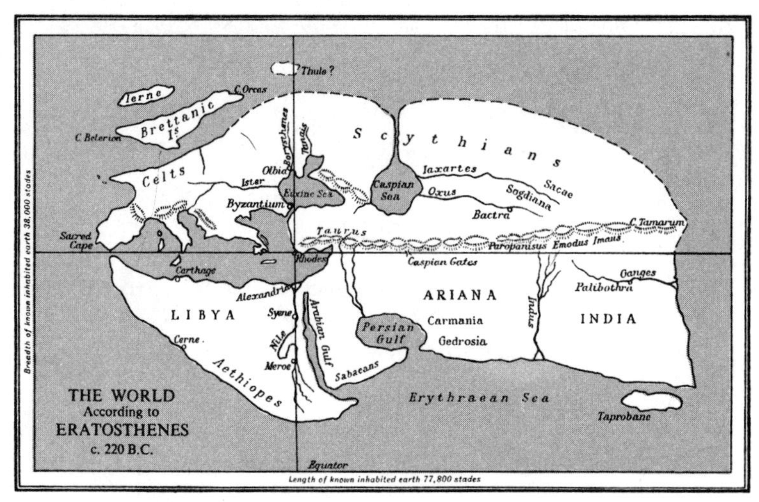

Fig. 4. Simplified reconstruction of Eratosthenes' map of the oikumene, based on north–south and east–west lines intersecting at Rhodes.

applied to land distances in Egypt, is that Eratosthenes may have used a stade of 157 m 50. This theory, well summarized by J. Oliver Thomson,[40] gives a much closer equivalent to the actual measurements, namely 39,690 km (24,663 miles).

Once this value of 252,000 stades was accepted, it was feasible also to work out the circumference of any parallel circle. Thus Eratosthenes calculated that the parallel of Rhodes, 36°N., was under 200,000 stades in circumference. To obtain the equivalent in stades of one degree of latitude he had only to divide by 360, i.e. 700 stades; to obtain the equivalent of one degree of longitude at Rhodes he could divide, say, 195,000 stades by 360, i.e. $541\frac{2}{3}$ stades. Thus there was the basis of fairly accurate co-ordinates for any sectional mapping of the Mediterranean based on the Rhodes parallel.

In his *Geographica* Eratosthenes discussed the best method of drawing a map of the inhabited area of the earth as known.[41] Then he calculated the distance along the Alexandrian meridian from the Cinnamon country in the south to Thule in the north as approximately 37,600 stades. The way he expresses this in detail is by distances north of the equator, of the tropic, and of various parallels; but in simplified form it works out as follows:

From	*To*	*Stades*
Equator	Cinnamon country	8800
Cinnamon country	Meroe	3000
Meroe	Syene	5000
Syene	Alexandria	5000
Alexandria	Rhodes	3750
Rhodes	Lysimachia	4250
Lysimachia	R. Borysthenes	5000
R. Borysthenes	Thule	16,500

His meridians, from west to east, may have been somewhat as follows:

Meridian	*Distance E. of the preceding in stades*
Western capes of Europe	—
Pillars of Hercules	3000
Straits of Messina/Carthage	8000
Rhodes/Alexandria	15,000
Issos	5000
Caspian Gates/Persian Gulf	10,000
R. Indus	14,000
Eastern frontier of India	16,000
Furthest capes of Asia	3000

As with earlier map construction, the length of the oikumene greatly exceeds the width, though by what proportion depends on how much of the northern, eastern and southern extremities was regarded as inhabited. It is clear from Strabo that Eratosthenes used an orthogonal projection. Rather than a rectangle, he thought of the oikumene as tapering off at each end of its length, like a *chlamys* (short Greek mantle). Moreover Strabo tells us that to the above total of 74,000 stades Eratosthenes, using another mathematical ploy, added 2000 at each end, to prevent the width being more than half the length. Germaine Aujac points out that on the parallel of Rhodes this total of 78,000 stades corresponds to about 140° longitude, which is roughly the distance from Korea to the west coast of Spain. But ancient methods of reckoning longitude were unreliable compared with those for latitude obtained from solar or stellar observations. Although in those times it was known that simultaneous eclipse observations could give a reliable result, how could scholars or practical seamen easily communicate and correlate such results?[42] On the whole the ancients used reported east–west land and sea measurements, and it was scarcely possible to keep these on the same parallel.

As approximations to sizes and shapes of parts of the world, Eratosthenes first divided the inhabited world by a line going from the Pillars of Hercules to the Taurus mountains and beyond, then subdivided each of these two sections into a number of irregular shapes. The word which the Alexandrian Greeks used for these was *sphragides*. *Sphragis* literally means 'a seal', especially an official seal; but the word was extended in the first place to a plot of land numbered by a government surveyor, then by extrapolation to a numbered area on a map. India he suggested drawing as a rhomboid, Ariana (the eastern part of the Persian Empire) as an approximate parallelogram. We do not know the total number of *sphragides* and have shapes recorded only for some.

This map of the known world was a very striking achievement and may be considered to be the first really scientific Greek map. Though we do not know its dimensions, as it was presented to the Egyptian court it may well have been fairly large. It must have been drawn as closely as possible to scale, and its influence on subsequent Greek and Roman cartography was tremendous. Indeed, with Ptolemy's inaccurate alterations to the overall dimensions of the world and the oikumene, it can be said to have affected world maps right down to the Age of Discovery.

THE SECOND AND FIRST CENTURIES BC

Under the influence of the so-called 'Scipionic Circle', a group of Roman philhellenes headed by Scipio Aemilianus, important cultural contacts between Greece and Rome, especially in the sphere of literature, were made. A Greek friend of Scipio, the historian Polybius (*c.* 200–after 118 BC), went on a voyage of exploration in the Atlantic after the destruction of Carthage in 146 BC. For his writings on geography, see p. 60. Like Eratosthenes, he maintained that the equatorial zone might be cooler than surrounding areas.[43] He added that it is very high, and therefore has a high rainfall, the clouds from the north causing precipitation at the time of the Etesian winds. (F. W. Walbank, in *A Commentary on Polybius* iii.576, rightly renders the superlative ὑψηλοτάτη thus, not as the highest point in the world.)

It is tempting to think of Polybius' treatise on the equatorial region as having in part been a criticism of the treatment of that area in a globe manufactured not long before. About 168 BC the Greek polymath Crates of Pergamum, who wrote among other things on Homer and the wanderings of Odysseus, visited Rome. He was professionally interested in the city's drainage system, but while exploring the Cloaca Maxima broke his leg. He used the period of recovery to give lectures in Rome, which are said to have created a great impression.[44] His view of terrestrial mapping was that the shape could only be right if it was drawn on a globe, and evidently that the scale could only be effective if the globe was at least 10 feet in diameter.[45] In designing his 'orb', if indeed he put his theory into practice, Crates favoured an unusual form of symmetry. There were, he said, separated by two intersecting belts of ocean, four symmetrical land-masses: (a) Europe, Asia and the part of Africa known at that time; (b) south of them, that of the Antoikoi, 'dwellers opposite'; (c) west of them, the Perioikoi, 'dwellers round'; (d) south of the Perioikoi, the Antipodes. The break between the land-mass known at that time and that of the

Fig. 5. Reconstruction of the inhabited area of Crates' Orb, a terrestrial globe made *c.* 170–160 BC to illustrate Homeric geography.

Antoikoi came, according to him, at a belt on each side of the equator, and there were Ethiopians (Aethiopes, 'black-faces') on each side of this water divide. Homer had written of

> the Ethiopians, split in two,
> Some in the East, some by the setting sun.[46]

Later Greek writers interpreted this passage in various ways.[47] No doubt as a Homeric scholar Crates was more concerned to give a plausible account of Homeric descriptions than to investigate explanations which suggested the existence of a continuous African land-mass stretching across the equator. The idea, however, was taken up by Cicero in the 'Dream of Scipio' (*somnium Scipionis*) which he incorporated in his *De re publica*. When Macrobius (p. 174 below) wrote a commentary on the *somnium Scipionis* about AD 390, he defended and amplified Crates' theory, aspects of which thus found their way into medieval cartography; the Perioikoi and Antipodes were then omitted, although discussed by Cicero and Macrobius.[48]

It remains to consider what criticisms Greek writers, in the follow-up to the Hellenistic Age, had to contribute to the theoretical side of terrestrial cartography. The two names which stand out in this connection are those of Hipparchus and Posidonius. Hipparchus of Nicaea (*fl.* 162–126 BC), who made astronomical observations in Rhodes, is chiefly known for his astronomical writings.[49] But some sixty-three fragments of his work *Against the 'Geography' of Eratosthenes* survive, mostly in Strabo. According to Strabo, Hipparchus was prepared to accept Eratosthenes' figure of 252,000 stades for the circumference of the earth, but claimed that many other distances quoted by him were either contradictory or mathematically impossible.[50] According to Pliny, on the other hand, Hipparchus added almost 26,000 stades to Eratosthenes' figure.[51] One theory is that he chose a figure between 252,000 and 300,000 stades, which latter Archimedes quoted as an earlier estimate;[52] another suggestion is that Pliny was confused by two measurements of Eratosthenes, one dividing the circumference at the equator into quarters, each measuring 63,000 stades,[53] the other estimating the breadth (north–south) of the oikumene as 38,000 stades.[54]

Strabo also criticizes Hipparchus' regional treatment. This concerns the distances in Eratosthenes' third *sphragis*,[55] covering the eastern part of the former Persian Empire. Here some, but not all, of Hipparchus' calculations are defended by Dicks against Strabo's allegations of manufacturing evidence.[56] From one fragment it is clear that Hipparchus followed the evidence presented by Pytheas for the sun's elevation and the maximum number of daylight hours in the north-west.[57]

Finally there is the significant, though misleading, criticism of Eratosthenes' measurement by Posidonius of Apamea (*c.* 115–51/50 BC), who lived mainly in Rhodes and among his many interests wrote on world geography. He studied Atlantic tides at Cadiz, was interested in geology, and supported Polybius on the equatorial zone (p. 61). He established that the star Canopus was on the horizon at Rhodes, while at Alexandria it reached an elevation equal to 1/48 of the great circle. He clearly considered Rhodes and Alexandria as being on the same longitude, but in fact Rhodes is 28°14′ E., Alexandria 29°51′ E. If the distance between the two was 5000 stades, as he first thought, the circumference of the earth would, on his wrong assumption about longitudes, be 240,000 stades. But later he accepted a revised estimate for Rhodes–Alexandria of 3750 stades, which reduced the figure to 180,000 stades. If the length of a stade was the same in his work as in Eratosthenes, which is not certain,[58] this would reduce the length of a degree of longitude at the equator from Eratosthenes' figure of 700 to 500 stades. Posidonius also believed that the east–west length of the known world was 180°. Both these suggestions, as will be seen, were taken up by Ptolemy, with unfortunate results because cartographers of a very much later period tended to take Ptolemy's word as law.

CHAPTER III

AGRIPPA

ROMAN MAPPING BEFORE AGRIPPA

There are only scanty records of Roman maps of the Republic. The earliest of which we hear, the Sardinia map of 174 BC, clearly had a strong pictorial element, and as such is discussed on p. 148 below. But there is some evidence that, as we should expect from a land-based and at that time well advanced agricultural people, subsequent mapping development before Julius Caesar was dominated by land survey; the earliest recorded Roman survey map is as early as 167–164 BC. If land survey did play such an important part, then these plans, being based on centuriation requirements and therefore square or rectangular, may have influenced the shape of smaller-scale maps. This shape was also one which suited the Roman habit of placing a large map on a wall of a temple or colonnade. Varro (116–27 BC) in his *De re rustica*, published in 37 BC, introduces the speakers meeting at the temple of Mother Earth (Tellus) as they look at 'Italy painted' (*Italiam pictam*). The context shows that he must be talking about a map, since he makes the philosopher among his group start with Eratosthenes' division of the world into North and South. This leads him on to the advantages of the northern half from the point of view of agriculture. The speakers compare Italy with Asia Minor, a country on similar latitudes where Greeks had experience of farming. After this they discuss in more detail the regions of Italy. As a visual aid to this discussion, the temple map will have been envisaged as particularly helpful. But whether it was only intended to be imagined by readers or was actually illustrated in the book is not clear. The same applies to possible cartographic illustration of Varro's *Antiquitates rerum humanarum et divinarum*, of which Books VIII–XIII dealt with Italy. But at least we know that he was keen on illustration, since his *Hebdomades vel de imaginibus*, a biographical work in fifteen books, was illustrated with as many as seven hundred portraits.[1] Since we are told that this work was widely circulated, some scholars have wondered whether Varro used some mechanical means of duplicating his miniatures; but educated slaves were plentiful, and we should almost certainly have heard about any such device if it had existed.

The only Roman world map before Agrippa's was the one which Julius Caesar commissioned but never lived to see completed. We are told by late Roman and medieval sources that he employed four Greeks, who started work on the map in 44 BC.[2] These were no doubt freedmen, of whom there were large numbers in Rome, including many skilled artisans. They are said to have divided the world for the purpose of this map as follows:

Region	*Cartographer*	*Years and months of work*	*Year of completion*
East	Nicodemus or Nicodoxus	21.5	30 BC
West	Didymus	26.3	27 BC
North	Theodotus or Theodocus	29.8	24 BC
South	Polyclitus	32.1	—

It will be seen that the combination of these figures[3] points to a date about nine or ten years earlier; but 44 BC, during Caesar's dictatorship, seems much more likely. The four regions of the world are not self-explanatory, but what Caesar seems to have meant is as follows: East, all to the east of Asia Minor; West, Europe except Greece, Macedonia and Thrace; North, Greece, Macedonia, Thrace and Asia Minor; South, all Africa. If Romans were planning this, they would place the northern section much further west, whereas these were Greeks, following a tradition which originated in Rhodes or Alexandria.

We may speculate whether this map was flat and circular, even though such a shape might have been considered 'unscientific' and poorly adapted to the shape of the known world. That is the form of the Hereford World Map (p. 179), which, however, seriously distorts the relative positions and sizes of areas of the world in a way we should not imagine Julius Caesar and his technicians would have; though the only artefact claimed as theirs, the Mauchamp stone (p. 102), must be admitted to be too uncertain evidence. A late Roman geographical manual gives totals of geographical features in this lost map without recording names, but even the totals turn out on examination to be unreliable.

AGRIPPA AND HIS MAP

A much quoted but unfortunately not extant map, that of Agrippa, was compiled to further Roman imperial expansion.[4] M. Vipsanius Agrippa (64/63–12 BC) was one of the earliest supporters of the young Octavian in his fight to establish himself as Julius Caesar's heir. He first became prominent as governor of Gaul, where he improved the road system and put down a rebellion in Aquitania. He pacified the area near Cologne (later founded as a Roman colony) by settling the Ubii at their request on the west bank of the Rhine. In 37 BC he was consul and built Octavian a fleet which enabled him the following year to defeat Sextus Pompeius in Sicily; Agrippa as admiral of this fleet used a new type of grapnel devised by him. His greatest victory was in 31 BC when off Actium, near Preveza in western Greece, Octavian and he defeated Antony and Cleopatra. He was one of the main helpers of Octavian when in 27 BC the latter was invested with special powers and the title Augustus. In 23 BC Augustus, as he was ill, handed his signet-ring to Agrippa, thus indicating him as acting emperor. The same year Agrippa was given charge of all the eastern parts of the Empire, with headquarters at Mitylene. In 21 BC he returned to Rome and married Augustus' daughter Julia. After he had put down the Cantabri of northern Spain in 19, he returned to Rome more permanently and was given additional favours. From 17/16 to 13 he was pacifying the eastern provinces, and in 12 BC went to Pannonia, but died shortly after his return.

Augustus had a practical interest in sponsoring the new map of the inhabited world entrusted to Agrippa. On the re-establishment of peace after the civil wars, he was determined on the one hand to found new colonies to provide land for discharged veterans, on the other hand to build up a new image of Rome as benevolent head of a vast empire. Mapping enabled him to carry out these objectives and to perfect a task begun by Julius Caesar. It became, among other things, a useful tool in the propaganda of imperial Rome. Agrippa was an obvious choice as composer of such a map, being a naval man who had travelled widely and had an interest in the technical side. He must have had plans drawn, and may even have devised and used large-scale maps to help him with the conversion of Lake Avernus and the Lacus Lucrinus into naval ports.

The world map, incomplete at Agrippa's death in 12 BC, was completed by Augustus himself. It was erected in Rome on the wall of a portico named after Agrippa, which extended along the east side of the Via Lata (modern Via del Corso). This portico, of which fragments have been found near Via del Tritone, was usually called Porticus Vipsania, but may have been the same as the one which Martial calls Porticus Europae, probably from a painting of Europa on its walls.[5] The building of this colonnade was undertaken by Agrippa's sister Vipsania Polla. The date at which the building was started is not known, but it was still incomplete in 7 BC. Whether the map was painted or engraved on the wall we do not know. The theory that it was circular must surely be wrong, as such a shape does not suit a colonnade wall: it is likely to have been rectangular, probably with north rather than south at the top.

The chief ancient writer who refers to it is the elder Pliny, who frequently quotes Agrippa by name; though whether in most cases his source is the map or the commentary is hard to say. Pliny's most specific reference to the map is in *NH* iii. 16–17, where he records that the length of Baetica, the southern Spanish province, was given as 475 Roman miles and its width as 258 Roman miles. Such measurements, he says, were obsolete by Agrippa's time. In fact the length by Augustus' time was about 280 Roman miles, whereas the width could still be correct, depending on how it was calculated. Pliny continues: 'Who would believe that Agrippa, a very careful man who took great pains over this work, should, when he was going to set up the map to be looked at by the people of Rome, have made this mistake, and how could Augustus have accepted it? For it was Augustus who, when Agrippa's sister had begun building the portico, carried through the scheme from the intention and notes [*commentarii*] of M. Agrippa.' In point of fact Augustus may have delegated the detailed checking to one of his freedmen, such as his librarian C. Iulius Hyginus.[6] Pliny's words do not make it clear whether the notes were written up as a separate commentary, and Detlefsen thought they were not.[7] Certain phrases in Pliny lead one to suppose they came from a commentary, not a map. Thus Agrippa is said to have written (p.

49 below) that the whole coast of the Caspian from the R. Casus consists of very high cliffs, which prevent landing for 425 miles. If the commentary had not been continuous, but had merely served as supplementary notes where required, there is a possibility that by Pliny's time, some eighty years later, it might have gone out of circulation. Two late geographical writings, the *Divisio orbis* and the *Dimensuratio provinciarum* (commonly abbreviated to *Divisio* and *Dimensuratio*) may be thought to come from Agrippa, because they show similarities with Pliny's figures.[8] But the numerals preserved in their manuscripts tend to be very corrupt. There are, however, cases, e.g. the combined measurements of Macedonia, Thrace and the Hellespont, which agree with Pliny[9] on areas where he does not name Agrippa but may nevertheless in fact have been using him. We may treat as secondary sources Orosius, *Historiae adversum paganos*,[10] and the Irish geographical writer Dicuil (*fl.* AD 825).[11] Orosius seems to have read, and followed fairly closely, both Agrippa and Pliny, as well as early writers from Eratosthenes onwards. Dicuil tells us that he followed Pliny except where he had reason to believe that Pliny was wrong.

It is also claimed that Strabo obtained his figures for Italy, Corsica, Sardinia and Sicily from Agrippa. His source was clearly one commissioned by Romans, not Greeks, as his figures for those areas are in miles, not stades. But Strabo never names this source, merely calling it 'the chorographer'. Such a word certainly ties up with *Divisio* 1: 'The world is divided up into three parts, named Europe, Asia, Libya or Africa. Augustus was the first to show it [the world] by chorography.' Evidently there is a slight difference of meaning between this and Ptolemy's definition, by which chorography refers to regional mapping. A direct comparison with Pliny, who may here be turning to Agrippa as his source although he does not name him, can be made out in the case of Sicily, where the figures do not tally:

Distances round Sicilian coast

STRABO		PLINY
15 legs	557 miles	
1 leg, estimated, S. coast	± 50 miles	
Total	± 607 miles	618 miles

Road distances

(a) Cape Pachynum–Cape Pelorus
168 miles 176 miles

(b) Messana (Messina)–Libybaeum (Marsala)
235 miles (MSS XXXV)
by Via Valeria 242 miles (MSS CXLII)

Unless Agrippa's map and his commentary gave different mileages, which is unlikely, the question arises whether Strabo went not to these, the most recent Roman source available to him, but to Julius Caesar's map.

For a more complete assessment of what Agrippa wrote or ordered to be put on his map, we may turn to passages where Pliny quotes him specifically as reference. These include both land and sea measurements, though the commonest are lengths and breadths of provinces or groups of provinces. In this context, 'length' normally means the greater of the two measurements. The fact that for continental measurements it also usually means west–east or north–west/south–east is largely coincidental.[12] Although the words used are *longitudo* and *latitudo*, they have no connection with longitudinal and latitudinal degree divisions.

Plin. *NH* iii.16. See p. 42 above. With reference to this length of Baetica, at iii.8 Pliny offers a possible explanation when he writes of Agrippa as considering that most of the coast of Baetica was earlier occupied by Phoenicians.

iii.37. 'Agrippa reports the length of Gallia Narbonensis as 370 miles, its width as 248.' The figures are corrected from Martianus Capella; similar ones are given in the *Dimensuratio*.

iii.86. 'The circumference of Sicily, according to Agrippa, is 618 miles.' See above; although Detlefsen wanted to emend the figure to 528, Dicuil has the same figure.

iii.96. 'The Lacinian promontory [C. delle Colonne, near Crotone, S. Italy] is reported by Agrippa to be 70 miles from Caulon [Kaulonia near Locri].' This is correct if understood as a straight sea measurement followed by the short distance up river.

iii.150. 'Illyrium has maximum width 325 miles, length from R. Arsia [Raša, Istria] to R. Drinium [Drin] 530. From the Drin to the Acroceraunian promontory [C. Gjuhëzës, Albania] was said by Agrippa to be 175 miles, while the whole Italian and Illyrian Gulf is

1700 miles round.' The latter measurement, for which, if individual measurements are added up, 1600 miles would be more correct, refers to the Adriatic from its head to C. Glossa and Otranto.

iv.45. 'From the mouth of the Danube to the outlet of the Black Sea some have reckoned as 500 miles; Agrippa added 60.' This translates *D. alii*, a very convincing emendation, rather than the MS reading *DLII*, which would clash even more with iv.78 (below).

iv.60. 'At its promontory Criumetopon ["Ram's forehead", now C. Krios], according to Agrippa, Crete is 125 miles from Cyrene's promontory Phycus (now Ras Sem, Libya). He also gives the distance from Crete at Mt Cadistus (NE spur of Mt Dikte) to Cape Malea in the Peloponnese as 80;[13] and from the island of Carpathos to Crete at the promontory of Samonium as 60 miles west.' Whereas Eratosthenes had reckoned the distance between Crete and Cyrenaica as 2000 stades (250 Roman miles), Pliny virtually goes back to ps.-Scylax (fourth century BC) who spoke of a twenty-four-hour sail.

iv.77. 'The distance round the Black Sea, according to Varro and most of the early writers, is 2150 miles; Cornelius Nepos adds 350, Artemidorus makes it 2119, Agrippa 2360, and Mucianus 2425.' Here Pliny has turned to more sources than usual, none in fact very early and ending with C. Licinius Mucianus, who had been governor of Syria in AD 68–69. The only difficulty with the figure given for Agrippa, 2360 miles, is that Pliny also attributes to him separate figures for the north and the south shore (see below), which when added together, even without the two straits involved, come to 2555 miles; so possibly 2560 was the intended number.

iv.78. 'Agrippa gives Byzantium–Danube as 560 miles and from the Danube to Panticapaeum [Kertsch] as 638 miles.' The MSS have *DCXXXV in*, but as the *in* is meaningless, *DCXXXVIII* was probably intended.

iv.81. 'Agrippa reported the whole area from the Danube to the ocean as 1200 miles in length, 396 in width, to the Vistula from the steppes of Sarmatia.' The width expressed (taking the reading of the second hand in two manuscripts) is literally 400 minus 4; an alternative width is the 386 miles given by the *Divisio* and *Dimensuratio*. The former mentions the area as Dacia, the latter as

Dacia and Getica. Pliny avoids these names, perhaps because Agrippa's map did not contain them. What route was reckoned from the mouth of the Danube to the Vistula is uncertain.

iv.83. '125 miles from Achilles' island is a sword-shaped peninsula called "Achilles' racecourse" [Dromos Achillis, the spit near the mouth of the Dnieper now known as Tenderovskaya Kosa], whose length Agrippa gives as 80 miles.' Strabo gives its length as 1000 stades, corresponding to 125 Roman miles.[14] That length was probably taken from Eratosthenes; between his time and Agrippa's the spit may well have been shortened by marine erosion; it is much shorter today.

iv.91. 'The length of Sarmatia, Scythia and Taurica and the whole stretch from the R. Borysthenes [Dnieper] is given by M. Agrippa as 980 miles, its width as 716 miles. I personally consider measurement in that part of the world uncertain.' Detlefsen suggested for *Scythiae Tauricae* the rendering 'Tauric Scythia';[15] but the Crimea, which is here referred to, is called by Pliny Taurica. The length of 980 miles could be thought to extend from the Don, since Asia is not here under consideration, to Hungary, to which detached groups of Scythians emigrated from the Crimea and south Russia from the fifth century BC. As to Scythia in Europe, Pliny comments (iv.81): 'The name of Scythians has altogether[16] been transferred to the Sarmatae and the Germans'. This shows that ancient map-reading, in marking in tribal names, had the problems of all political maps. The length given by the *Divisio* and the *Dimensuratio* is also 980 miles, so that this rather than the alternative 880 given by some Pliny manuscripts should be preferred.

iv.98. 'The Greeks and some of our writers have given the coast of Germany as 2500 miles long. Agrippa gives the length of Germany, Raetia and Noricum as 686 miles, width 248; whereas in fact the width of Raetia alone is perhaps longer than that, though admittedly it was conquered about the time of his death.' An alternative reading for the length is 636; the *Divisio* gives about 800, the *Dimensuratio* 623. Pliny's criticism of the length reported by Agrippa is fully justified.

iv.102. 'Agrippa believes the length of Britain to be 800 miles, its width 300; the same width for Ireland, but its length 200 miles less.' Agrippa agreed with Julius Caesar's measurements quoted from early writers for the length of Britain, 800 miles, but not with

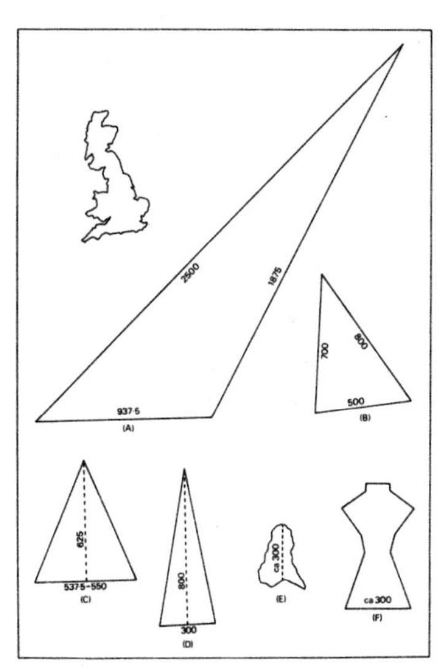

Measurements in Roman miles

Fig. 6. Shapes and sizes of Britain in ancient writers: (A) Diodorus Siculus, from Pytheas or Eratosthenes; (B) Caesar; (C) Strabo; (D) Agrippa; (E) Mela; (F) Tacitus.

Caesar's fairly accurate south coast figure of 500, unless indeed Pliny's width of 300 miles exceptionally denotes an average.

iv.105. 'Agrippa reckoned the coastline of the provinces of Gaul as 1750 miles, and the length of those provinces between the Rhine and the Pyrenees and the ocean and the Cevennes and the Jura, excluding Narbonese Gaul, as 420 miles, their width as 318 miles.' As with Germany, Agrippa's length and breadth are too short. Dicuil, who announced in his preface that he would emend Pliny's figures if they seemed wrong, corrected 420 to 920; the *Divisio* has 828 for the length.

iv.118. 'Agrippa made Lusitania, together with Asturia and Gallaecia, extend in length for 540 miles, in width for 536.' The measurements given in *Divisio* are 480, 450; in *Dimensuratio* 580, 585.

v.9–10. 'Polybius . . . says that west of Mt Atlas are woods and upland pastures [*saltus*] full of wild African animals as far as the R. Anatis [in Mauretania, south of the Roman province], 496 miles. From there to Lixus [Larache] Agrippa makes 205 miles, saying that Lixus is 112 miles from the Straits of Gibraltar. From there he lists a gulf called Sagigi, a town on the promontory of Mulelacha, the rivers Sububa and Salat, the harbour of Rutubis [= Rusibis, probably the modern Mazagan] 234 miles from Lixus, then the promontory of the Sun [Solis, but probably adapted from Greek Soloeis], the harbour of Rhysaddir [perhaps near C. Ghir], the Gaetuli Autoleles [or Autoteles], the R. Quosenus [Sous], the tribes Velatiti and Masati, the R. Masathat [Massa], the R. Daret [Dra], in which are crocodiles. Then, he says, a gulf 616 miles across is shut in by a promontory of Mt Braca, projecting to the west, which is called Surrentium. After this is the Salt River, beyond which are the Ethiopian Perorsi, and behind them the Pharusii. Next to these inland, according to Agrippa, are the Gaetulian Darae, but on the coast Ethiopian Daratitae and the R. Bambotus, full of crocodiles and hippopotami. From there, according to him, are continuous mountains up to the one called Theōn Ochēma, "Chariot of the Gods", which we shall mention. From that mountain to the promontorium Hesperium, "Western Cape", is a sail of ten nights and days. In the midst of that space he placed Mt Atlas, which all others give as being in the furthest areas of Mauretania.'

This passage is fully discussed by J. Desanges in his Budé edition of the African section of Book V of Pliny's *Natural History*.[17] He thinks that most of the coastal passage, from R. Anatis southward, is taken from Agrippa, as indicated by the reported speech construction, but that Theōn Ochēma and the Western Cape are inverted, because Pliny took them from Hanno's voyage. Whereas the distance is here given as ten days and nights, in *NH* vi.197 it is given as only four days' sail. Clearly these were relatively unexplored areas. Theōn Ochēma (p. 132) is most generally equated with Mt Kakulima, near Conakry, Guinea. The figure of 616 miles corresponds to eleven days' sail at fifty-six miles a day, a figure adapted from Polybius. The R. Bambotus is variously explained as associated with a modern regional name Bambouk in Upper Senegal or from the rare Greek compound *pambotos*, 'all-nourishing'.[18] Most of the places mentioned lack measurements, and where they are ascertainable, e.g. Lixus, they are incorrect.

v.40. 'Agrippa gives the length for the whole of Africa from the Atlantic, including Lower Egypt, as 3080 miles.' This figure is an emendation by Desanges for $\overline{LXXX}$ (= 80 miles) in the MSS: Martianus Capella gives 3040.

v.65. 'Agrippa gives the distance from Pelusium [north-east of el-Qantara, Egypt] to Arsinoe [near Suez], a town on the Red Sea, through the desert as 125 miles. Such a small distance is it between two such differing regions.' Much earlier, Herodotus had given the same distance, 1000 stades.[19] The ship canal from the Mediterranean to the Red Sea, dug by Pharaoh Neco, was renewed by Darius I, Ptolemy I and II and Trajan.

v.102. 'Agrippa divided Asia into two parts. One he delimited by Phrygia and Lycaonia on the east, by the Aegean on the west, by the Egyptian Sea on the south and by Paphlagonia on the north; he made its length 470 miles, its width 320. The second he delimited by Armenia Minor on the east, Phrygia, Lycaonia and Pamphylia on the west, on the north the province of Pontus, on the south the Pamphylian Sea: 575 miles long, 325 miles wide.' The areas in question are both parts of Asia Minor, the first being the western half. Although the province of Pontus is mentioned, the boundaries are not intended to follow those of Roman provinces. For further Asia see on vi.37 and 57 below.

vi.3. 'Agrippa gives the distance from Chalcedon to the Phasis [i.e. the Bosporus plus the distance to the furthest place east on the Black Sea] as 1000 miles, and from there to the Cimmerian Bosporus [area east of the Sea of Azov] as 360 miles.' For the distance round the Black Sea see on iv.77. Agrippa's measurement of 1000 Roman miles (about 1500 km) is very generous and must include indentations; it is taken over from Eratosthenes.[20]

vi.37. 'Agrippa states that the Caspian, with the tribes round it, and Armenia, delimited on the east by the China Sea, on the west by the Caucasus, on the south by the Taurus, and on the north by the Scythian Ocean, extend as far as is known 480 miles in length, 290 in width.' These measurements, like those in many other ancient writers on the East, are absurdly small. The phrase used for the China Sea is Oceanus Sericus.[21] The length of 480 Roman miles is confirmed by the *Divisio* and the *Dimensuratio*.

vi.39. 'Agrippa writes that the whole shore of the Caspian from R. Casus, with very high rocks, has no access for 425 miles.' The shore described is either the west or the south; the R. Casus is not mentioned elsewhere. It seems likely that Agrippa thought of the Caspian as having an outlet northwards.

vi.57. 'Agrippa gives the length of India as 3300 miles, its width as 1300.' One manuscript gives the width as 2300, which sounds more reasonable. Artemidorus gave 2000 miles between the Indus and the Ganges; Agrippa's extra figure is intended to apply the term 'India' to the whole area as far as the Pacific (*Oceanus Eous*,

'Eastern Ocean', is the phrase in the *Dimensuratio*).

vi.137. 'Agrippa stated that Media, Parthia and Persis, delimited on the east by the Indus, on the west by the Tigris, on the north by the Taurus and the Caucasus, on the south by the Red Sea [i.e. Persian Gulf], extend in length 1320 miles, in width 840; also that Mesopotamia by itself, delimited by the Tigris on the east, the Euphrates on the west, the Taurus on the north and the Persian Sea on the south, is 800 miles long and 360 wide.' The width agrees with, and the length comes close to measurements given in vi.126. Agrippa's original may have had measurements in *schoinoi* (1 *schoinos* = 40 stades according to Eratosthenes, though other equivalents are found).

vi.139. 'Charax was previously 10 stades [= 1¼ Roman miles] from the coast, and Agrippa's portico too has it by the sea. Juba gave 50 stades, and now [*c.* AD 70] it is said to be 120 stades [= 15 Roman miles] from the coast.' This is a specific reference to the map on the portico wall in Rome rather than to the commentary. Charax was at the confluence of the Tigris and the R. Karun. The geographer Isidore (p. 124), probably of the Augustan Age, who was one of Pliny's sources, came from there. Juba II, king of Mauretania 25 BC–*c.* AD 23 and thus outliving Agrippa, gave Charax as the equivalent of 6¼ Roman miles from the sea. On a relatively small-scale world map one could not tell the difference between a seaside place and one 1¼ or 6¼ miles inland. Today the Abadan-Khorramshahr area is about 70 km (47 Roman miles for comparison) from the sea. But perhaps not too much faith should be put in ancient comments about the rate of coastal changes or river deposit. The chief interest of the Pliny quotation may be said to be the direct reference to the consultation of a map.

vi.164. 'Agrippa gave the length of the Sinus Arabicus [Red Sea] as 1732 miles,[22] each side having the same length.' Klotz thinks there may have been an adaptation of Eratosthenes' figures,[23] which seem to have been 14,000 stades for the Arabian coast, 13,500 for the African side, totalling 27,500. By the current Roman equivalent,[24] this works out at 1718¾ Roman miles for each side if they were equal. But as in west Africa, the measurement may be based on average length of day's sail: sailing for 31 days at 56 Roman miles a day would give 1736 miles, from which 4 may have been subtracted for the supposed distance of Arsinoe from the head of the Red Sea.

vi.197. 'Agrippa thought the whole land of the Ethiopians, including the Red Sea, extended in length 2170 miles; in width, including Upper Egypt, 1296 miles.' Here *mare Rubrum* was seen to mean the Red Sea, whereas in passages quoted above it means the Persian Gulf and *sinus Arabicus* the Red Sea. How the figures were arrived at is uncertain; the areas said to have been occupied by Ethiopians vary enormously between ancient authors.

vi.206–7. 'Polybius gave the distance from the Straits of Gibraltar direct east to Sicily as 1250 miles, Sicily–Crete 375, Crete–Rhodes 187½, Rhodes–Chelidoniae (Şelidanburu, five rocky islands off the Turkish coast), the same, Chelidoniae–Cyprus 225, Cyprus–Seleucia Pieria in Syria 115, total 2340 miles. Agrippa reckons this same distance from the Pillars of Hercules [Straits of Gibraltar] to the Gulf of Issus [Gulf of Iskenderun] in a straight line as 3440 miles. But there may be a miscalculation here, since he also gave the distance from the Straits of Messina to Alexandria as 1350 miles.' Clearly, if Pliny's statement is correct, Agrippa made a serious overestimate. Pliny's explanation seems to be that Agrippa, despite his wording, really calculated the distance via Alexandria and thence round the coast, perhaps basing his figure on a map with indentations. Even so, this figure of 1350 cannot possibly have been calculated 'in a straight line'; Riese suggested that it appeared in the commentary, not on the map.[25] Might one not suggest rather that someone adding up stretches for Agrippa inadvertently wrote the total as 3440 instead of 2440, or that Pliny was consulting a copy of the commentary which at this point was corrupt?

vi.208–9. 'The length of Africa, taking the average of various estimates, comes to 3798 miles. Its width, to the extent that it is inhabited, nowhere exceeds 750 miles. But as in the Cyrenaic part of Africa Agrippa made the width 910, including its deserts, where known, as far as the Garamantes, the whole measurement to be calculated comes to 4608 miles.' The length is east–west, whereas the 910 miles are a north–south measurement, from the Mediterranean to the Garamantes in the Sahara; so these two should not be added up. In any case either Pliny has added incorrectly[26] or the manuscripts are corrupt. In *NH* v.38 he says that the width of Cyrenaica, as far as it is known, is 810 miles; so probably 910 is here a mistake for 810. Contrary to Pliny's curious attempt to add to the length, Agrippa's own figure for the length, according to Pliny, *NH* v.40 (p. 48 above), was the much lower one of 3080 miles.

Book VII of the *Natural History* is anthropological, and in it Agrippa's commentary does not appear, so that it is unlikely to have had any concern with descriptions of tribes, which perhaps helps to confirm its concentration on mapping.

It is a pity that Pliny, who seems to be chiefly interested in measurements, gives us so little other information about Agrippa's map. For a general description, however, of what is meant by chorography we may turn to Strabo ii.5.17 (as mentioned above, Strabo nowhere names Agrippa): 'It is the sea above all which shapes and defines the land, fashioning gulfs, oceans and straits, and likewise isthmuses, peninsulas and promontories. But rivers and mountains too help with this. It is through such features that continents, nations, favourable sites of cities, and other refinements have been conceived, features of which a regional [chorographic] map is full; one also finds a quantity of islands scattered over the seas and along the coasts.' Clearly Agrippa's map had many of the above features, but whether it also contained main roads is uncertain. The only specific reference, in connection with Sicily (p. 43), is to Strabo's 'chorographer', who is not very likely to have been Agrippa.[27]

Although the term *chorographia* literally means 'regional topography', it seems to include fairly detailed cartography of the known world. The map probably did not, in the absence of any mention, use any system of latitude and longitude. It no doubt inherited a system of regional shapes from Eratosthenes. It is, as one might expect, more accurate in well-known than less-known parts, and more accurate for land than for sea areas. From the above quotations there would appear to be a general tendency to underestimate land distances in Gaul and Germany and in the Far East, and to overestimate sea distances. If west Africa is any guide, in areas where distances were not established they were probably entered only very selectively. What purpose was served by giving a width for the long strip from the Black Sea to the Baltic is not clear.

But on the credit side, Agrippa's map, sponsored by Augustus, was obviously an improvement on that of Julius Caesar on which it is likely to have been based. The fact that such an insignificant and distant place as Charax was named on the map shows the detail which it embodied. Moreover it seems to have been the first Latin map to be accompanied by notes or commentary. Romans going to colonies, particularly outside Italy, could obtain information about the location or characteristics of a particular place. Also the full extent of the Roman Empire could be seen at a glance.

MAPS IN LATIN POETRY

We do not hear of any maps in the works of Virgil and Horace, though Horace's journey to Brundisium (*Sat.* i.5) at times reads like an itinerary. From Ovid and Propertius, however, we have short extracts showing that they had become familiar in everyday life. Ovid, in *Heroides* i.5, reconstructs a letter which Penelope might have sent to the long-absent Ulysses. At the same time he makes his warriors, as they return home after ten years of fighting, behave like contemporary Romans. One of these is described as sketching the plain of Troy in wine on the table:

> Here flowed the Simois, there Sigeum's land;
> Here Priam's lofty palace once did stand,

and goes on to locate the tents of Achilles and Ulysses and the place where Hector's mangled body was dragged round.

Propertius makes his Arethusa lament that she has, in Lycotas' absence, to spend the winter nights learning about the course of the R. Araxes in Armenia, where he is campaigning, and how many miles Parthian horses gallop without any water, and that she is compelled 'from the map [*tabula*] to learn up painted worlds.'[28]

VITRUVIUS

Although the *De architectura* of Vitruvius (*fl.* 30–20 BC) contained diagrams, only one is preserved in the oldest manuscript,[29] namely a windrose which Vitruvius says 'is so drawn that it is clear where the winds come from.'[30] In another passage he writes of the sources of rivers 'painted or written in all over the world by chorographers'.[31]

THE AUTUN WALL-MAP

The tradition of Roman wall-maps continued in two forms: the city plan (pp. 103ff.) and the world map. No examples of the latter survive, but we have a description of the educational value of such a map. The rhetorician Eumenius, born *c.* AD 264, writes of a map which he is planning for the school at Augustodunum (Autun, France): 'Also let the schoolchildren see it in those porticoes and look every day at all lands and seas and every city, race or tribe that unconquerable emperors either assist by their sense of duty or conquer by their valour or control by inspiring fear. There you have seen, for educational purposes, visual aids to supplement what is difficult to absorb merely from being told about the situations, areas and distances of all places. Their names are written in, the sources and mouths of all rivers are indicated, as are all coastal indentations, and the parts where the ocean either encircles the world or makes an inroad into land-masses.'[32] The word rendered 'world' is *orbis* (= *orbis terrarum*), but despite the literal sense 'circle' this does not indicate that the proposed map was to be circular. As with Agrippa's map, its position on a portico wall suggests a rectangular shape.

CHAPTER IV

GEOGRAPHICAL WRITERS

The Greek and Latin writers to be examined here, who include poets, historians and encyclopaedists as well as geographers, were not noted as map-makers, but from Hecataeus onwards they may have incorporated some maps in their works. They certainly used them, and many are likely directly or indirectly to have inspired map-makers.

HOMER

Early Greek literature was exclusively poetry, and the first 'geographical' poem, if we may consider it as that, was Homer's *Odyssey*. It is usually held today that the *Iliad* and *Odyssey* were composed, probably orally in the first place, in the eighth century BC, but that they reflect traditions of many centuries earlier. The birthplace of Homer was contested even in antiquity, but was certainly on the west coast of Asia Minor or an adjoining island. The poet shows himself (or herself according to Samuel Butler[1]) fairly knowledgeable about the Peloponnese and the islands to the west of it, Odysseus' Ithaca being almost certainly Ithaki;[2] far less so about those regions more distant from his area. Ancient scholars vied with each other in locating these, as did Apollonius Rhodius (p. 31). It was generally agreed that the Lotus Eaters were in north Africa, probably on the island of Djerba; that the Cyclops lived near Mt Etna; that Scylla and Charybdis were one on each side of the Straits of Messina; and that the island of Phaeacia was Corfu.

The question whether Homer used any maps depends not on the topography of the *Odyssey* but on the description in the *Iliad* of Achilles' shield, discussed elsewhere.[3] This should probably not be thought of as a map, more as a complex work of art whose poet (*poietes* = 'maker') seems to have seen a rudimentary map. True, the description ends:

> On it he placed the mighty strength of Ocean,
> Beside the well-made buckler's outer edge.[4]

From this one might imagine that there was some attempt at reality, since the river Oceanus surrounding the inhabited earth was a permanent concept in Graeco-Roman antiquity. But the rest of the description does not tally: there is first a celestial section, then two cities, described in great detail, as mentioned, with a surrounding countryside, town and country activities both being outlined. One can perhaps envisage the terrestrial features as somewhat resembling the Thera (Santorin) fresco mentioned on p. 17 above.

CLASSICAL GREECE

The first Greek geographical work, in prose, was the *Periegesis* ('guide round') or *Periodos Gēs* ('journey round the world') of Hecataeus of Miletus, written about 500 BC. It was in two books, 'Europe' and 'Asia', the latter also including Africa. Hecataeus had travelled in Asia and Egypt, and is said to have made a map, well spoken of by Agathemerus[5] and probably based on that of Anaximander of Miletus (p. 22). Modern reconstructions reasonably show this map, which is likely to have accompanied his book, as circular, with the Ocean surrounding a land-mass which ends eastwards at the R. Indus and the Caspian 'Gulf'. The circular

Fig. 7. Conjectural reconstruction of Hecataeus' map of the oikumene, *c.* 500 BC. The Tin Islands, Cassiterides, are variously placed off Britain or elsewhere in the north-west.

shape of flat maps in the early period was laughed at by later Greeks. Hecataeus is the first writer we know of to think of the Caspian as flowing into the Ocean, an idea which long persisted. On the Black Sea, an area colonized by Miletus, he was clearly informative, and his writing may have helped Herodotus on areas like Thrace. To the north of the Danube, according to Hecataeus, were the Rhipaean ('gusty') Mountains, beyond which were the Hyperboreans. For Classical Greeks these were 'men of the far north'; Herodotus says: 'If there are Hyperboreans, there must also be Hypernotians ["men of the far south"].[6] Ancient and modern opinion has differed on the location of the Rhipaean Mountains; Aeschylus and Pindar imagined that they were at the source of the Danube. In west Africa Hecataeus clearly knew some places on the coast of Morocco. His view of the Nile seems to have been that it came, somehow, from the southern ocean.

A more critical approach to geography and cartography is to be seen in Herodotus (*c.* 495–*c.* 425 BC). He, like Hellanicus (*c.* 490–*c.* 405 BC), was not so much a geographical writer as a historian with a strong interest in geography. The works of Hellanicus, who was too prolific a writer, have not survived. He devoted whole works to regional history or ethnography, and did not use maps.[7] Herodotus, as has been seen (p. 24), despised all existing maps, and we have no evidence that he inserted any. Yet his ideas may well have influenced the development of Greek cartography.

Since he believed that a description of the background to the war between Persia and Greece was as important as the history of that war itself, Herodotus devotes the first half of his work to a full account of the Persian Empire, its origins, customs, institutions and wars. His account of world geography[8] comes as a digression after his remark about the Hyperboreans mentioned above. The maritime expeditions which he describes are that organized by Darius from the Indus to the Red Sea; that of Pharaoh Neco (p. 24); and that of the ill-fated Sataspes, executed by Xerxes because he turned back to the Mediterranean from the west coast of Africa instead of circumnavigating the continent, saying that his ship could go no further: it was 'stopped'. Unfortunately, Herodotus is more interested in personalities than in the details of these voyages. His accounts of the time taken, which relate to the entire journeys, show how slow and difficult they were. He cannot guess why three different names, Europe, Libya, Asia, have been given to the earth,

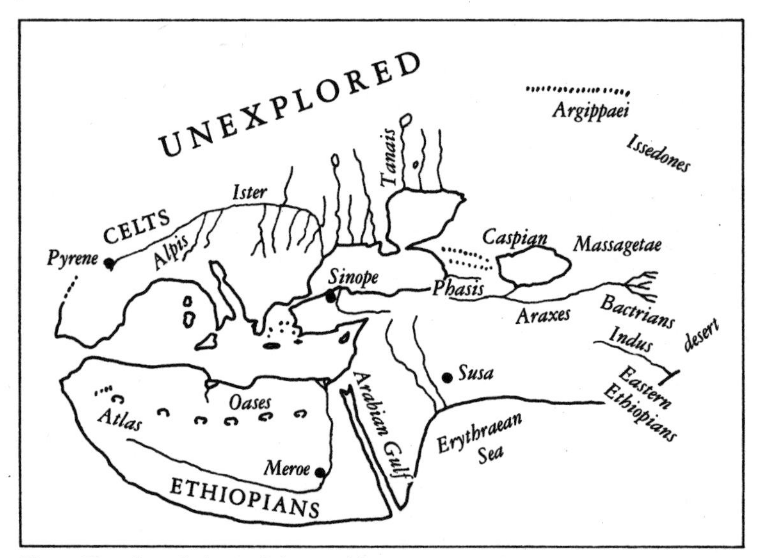

Fig. 8. Herodotus' view of the world, *c.* 450 BC.

which is one unit; why they are, according to him, named after women;[9] nor who fixed the boundary of Asia and Africa at the Nile and the boundary of Asia and Europe at the Colchian Phasis (or at the Don and the Straits of Kertsch). He does not know whether Europe is surrounded by water to west and north,[10] nor where the Cassiterides Islands are from which tin is obtained;[11] and the R. Eridanus, later equated with the Po but thought to come from amber-producing areas, may not, according to Herodotus, exist. Libya (Africa), he says, can be circumnavigated, as shown by the Phoenicians commissioned by Pharaoh Neco.[12] Asia as far as the Indus, he tells us, was explored by King Darius. But in no case does he give much detail.

As against this somewhat negative approach, Herodotus has a keen interest in such features as geomorphological change. He says that the area between the mountains beyond Memphis (Gizeh, near Cairo) seems at one time to have been a bay of the sea, as were the plains of Troy, Teuthrania (near Pergamum), Ephesus and the Maeander (Menderes), all of which he attributes to river deposit; while the R. Achelous in western Greece has linked islands to the mainland.[13] The choice of the Maeander is interesting because through later Greek usage[14] it led to the English 'meander'. Herodotus continues: 'There is also in Arabia near Egypt a bay of the sea [Gulf of Aqaba] whose length is forty days' rowing and its width half a day's. In it there is a daily tide,[15] and I think Egypt was at one time a similar bay. . . . If the Nile were to turn its stream into this Arabian Gulf, what would prevent it from being filled with soil within 20,000 years? Personally I think it would be within 10,000.' As evidence for his theory he quotes a record that nine hundred years earlier, when the Nile rose 8 cubits, it flooded all Egypt below Memphis; whereas in his day, unless it rose 15–16 cubits, it would not do so. This eventually leads him on[16] to the cause of the annual flooding of the Nile and to the location of its source. The treasurer of the temple at Sais told him the source was on two mountains between Syene (Aswan) and Elephantine. But he first traces it much further south, past Meroe to the land of the Automoli, then westwards, on the basis of explorations into the Sahara by some young Nasamonians reported by the priests of the oracle of Ammon (Siwa oasis). This leads Herodotus to conjecture that the Nile is roughly symmetrical with the Danube, also rising far to the west.

Democritus of Abdera in Thrace, who is best known as founder with his master Leucippus of the atomic theory, also wrote a *Geography* now lost. He is said to have been the first to suggest that the known world was roughly rectangular, with proportions, as mentioned (p. 25), east–west 3: north–south 2. This extent east–west implied ignorance of the Far East, but it was accepted as correct, e.g. by Dicaearchus, writing as late as a century later.[17]

THE PERIOD OF ALEXANDER'S EXPLORATIONS

Although the main objective of Alexander the Great was to conquer the Persian Empire, he had also a serious scientific project in mind. In the first place he arranged in advance for the compilation of a large body of information on the geography and ethnography of the regions likely to be penetrated. Then he took with him scientists in various disciplines, so as to have very full documentation compiled in the field (p. 29). After conquering the Persians, Alexander wanted to advance as far east in Asia as he could, but his army revolted. He compromised by ordering his admiral Nearchus to sail back via the mouth of the Indus, and he himself marched along the Baluchistan coast under great difficulties, caused chiefly by intense heat, mountainous sand-dunes and harassment by tribesmen. Had he not died in Babylon, he might well have carried out a plan to explore the western regions of the known world. It was rumoured that his last plan of conquest was to march west along the north African coast, besiege Carthage, and carry on as far as the Straits of Gibraltar.

The Greeks necessarily explored sea routes so as to keep up trade with their very widespread maritime colonies. One of the more distant of these in the West was Marseilles (Massilia), founded by Phocaea about 600 BC. About 320 BC, very soon after the death of Alexander, Pytheas of Massilia set out by sea to explore some of the northern waters of Europe. Since his voyages can be regarded as *periploi*, they are described on pp. 136f.

POLYBIUS

A Greek who formed over many years a very strong link with Rome was the historian and statesman Polybius (*c.* 200–after 118 BC). He became a friend of Scipio Aemilianus and was present with him at the destruction of Carthage in 146 BC. Shortly before this, the preoccupation of Carthage with its own defence opened up new routes for Greeks and Romans. The elder Pliny writes, 'When Scipio Aemilianus was in command in Africa, the historian Polybius was provided by him with a fleet for exploring that continent, and sailed round (i.e. explored NW Africa by sea). He reported that from Mt Atlas to the west, for 496 miles to the R. Anatis, there are *saltus* (upland woods with clearings) full of the wild animals that Africa produces'.[18] Polybius devoted Book XXXIV of his *Histories* to geography; it does not survive, except for many references in Strabo and a few quotations from other authors. Some of these give measurements, some describe coasts, rivers, springs, passes (he knew Hannibal's route at first hand), volcanoes; but the majority are of no cartographic interest. His comment on geographical writers is that he will ignore earlier ones, but examine critics of them such as Dicaearchus, Eratosthenes, 'the latest writer on geography', and Pytheas, who 'has misled many by saying that he traversed the whole of Britain on foot, giving the island a circumference of more than 40,000 stades, and telling us about Thule too.'[19] However, Strabo criticizes Polybius' measurements, maintaining that, although in the Mediterranean Polybius claims he is correcting Dicaearchus and Eratosthenes, he sometimes makes worse mistakes himself. He made the distance from Cape Malea (S. Greece) to the Danube 10,000 stades,[20] but according to Artemidorus the correct distance is 6500: 'the reason for this is that Polybius did not measure the distance in a straight line but by a chance route taken by some general'.

This Artemidorus (*fl.* 104–101 BC), a native of Ephesus, wrote eleven books of geography and *periploi*, which have not survived. He travelled extensively and made reasonable calculations of distance, though he sometimes misapplied Roman sources.

As to Posidonius, whose measurement of the circumference of the earth has been mentioned (p. 38), we are told that he constructed a globe and drew a map. The fragments of his history in fifty-two books are meagre; but Strabo quotes extensively from a separate work of his, *On The Ocean*.[21] Reviewing theories of the division of the earth into five zones,[22] Posidonius said that if the approach to this division were to be celestial, there should be new names, one central zone to be named amphiscian (with shadow coming sometimes from north, sometimes from south); then two heteroscian (shadow coming from one direction in the northern, from the opposite in the southern hemisphere; then two polar zones to be called periscian (with encircling shadow). But if human life were uppermost, one should add two narrow zones beneath the celestial tropics and crossed by the terrestrial tropics; these he considered the hottest areas of the world.

This must have been in an introductory section not confined to the ocean. The main novelty in the section on the ocean itself concerned the explorations of Eudoxus of Cyzicus.[23] Some time between 146 and 116 BC this Eudoxus was sent by Ptolemy Euergetes II, King of Egypt, to explore the sea route to India, and again by Cleopatra III after 116. On his return journey he was driven south of Ethiopia and made notes on a native language. He also found a horse figurehead which he later proved to have come from Gades (Cadiz), some ships' captains recognizing it as having belonged to a ship that sailed further south than was intended on the west coast of Africa. This caused him to make repeated attempts to sail round Africa to India from Gades; when he was exploring with his third ship, he recognized words from the east coast language. Eventually all trace of him was lost, and Strabo even disbelieves the whole story.

STRABO

We are fortunate in possessing all seventeen books of the *Geography* of Strabo,[24] written in good Greek although he himself was of mixed Asiatic and Greek stock; it is through his writings that most of our knowledge of Eratosthenes' mapping has come down. He was born at Amasia in Pontus in 64 or 63 BC; his grandfather had been an opponent of Mithridates and was considered by some a traitor as he handed over fortresses to Rome. Strabo was educated at Nysa near Tralles in Caria and in 44 BC went to Rome, where he studied under the Phoenician freedman Tyrannio[25] and the Stoic philosopher Athenodorus. He showed himself a keen supporter of Augustus and the *pax Augusta*, and visited Rome several times. From about 25 to 20 BC he was in Egypt, based at Alexandria. His *Geographica* was written between 9 and 5 BC and parts revised in AD 18–19. It seems not to have been read in Rome in the first century, not even being mentioned by the elder Pliny. Strabo travelled widely and lived to at least AD 21.

He is a lengthy and discursive writer, but shows good critical power in assessing earlier geographical writers and giving us a verbal picture of the known world of the time. He treats Homer as the first writer on geography, and defends the Homeric picture of the known world as substantially true. But within the Homeric chapters he has a section (i.1.8–9) in which he attempts to analyse navigation of the oceans over the ages. Thus he says: 'It is not reasonable to suppose that the Atlantic consists of two seas, confined by narrow isthmuses so as to prevent circumnavigation; rather it must be confluent and continuous.' His argument is that explorers tried to sail round Africa but turned back when not obstructed by any land-mass. The problems of the armchair geographer are revealed in the journalistic trick of quotes from quotes on an important exploration: 'Posidonius says Herodotus thinks that certain men sent by Neco completed the circumnavigation.'[26] This is all he reports, so that we have to beware of using all his work as scientifically worthy material. Perhaps because he is drawing on an account at second hand, he is afraid to support what may have seemed like science fiction. He does not deal extensively with Hanno the Carthaginian (p. 132), instead spending much effort on questioning the explorations of Eudoxus of Cyzicus mentioned above, who must in fact have added to the accumulation of knowledge about the remoter parts of Africa.

Strabo likes to represent myths and poetic phraseology geographically. Thus he says[27] that the legend of the Golden Fleece brought back from Colchis by the Argonauts reflects the search for gold by early Greeks in areas of the Black Sea.[28] When Homer made Hera say:

> For I shall see the bounds of fertile earth
> And Oceanus, father of the gods,[29]

what he means, says Strabo (i.1.7), is that Ocean touches all the extremities of the land. Or again (i.1.20), when Homer describes Odysseus as seeing land as he was on the crest of a great wave,[30] he must have been referring to the curvature of the earth, a phenomenon familiar to sailors. Some of this was polemic against Eratosthenes, who would not have it that early epic poetry could contribute anything to scientific theory.

The most detailed examination of a term arising from Homeric geography is in respect of Ethiopians,[31] a problem already discussed in connection with Crates (p. 37). What did Homer mean by saying they were 'divided in two, some where Hyperion rises and some where he sets'?[32] The historian Ephorus (*c.* 405–330 BC) mentioned an early tradition that Ethiopians had overrun Libya, i.e. north Africa, as far as Dyris (the Atlas mountains), and that some had stayed there. Crates' view was based on an unorthodox view that the division was north–south rather than the obvious interpretation of east–west. Aristarchus of Samothrace (*fl.* *c.* 155 BC) criticized Crates' interpretation, but claimed that Homer was simply wrong and there was only one area in which Ethiopians lived. Strabo's own view is that there were two groups of Ethiopians, one living in Asia and one in Africa;[33] and that Homer thought likewise, though not to the extent of placing the eastern group in India, of which he had no knowledge. However, this idea of eastern Ethopians living in some area of India and resembling Indians in appearance and customs persisted throughout antiquity.[34]

The function of geography, according to Strabo, is to be an interpreter not of the whole world but of the inhabited world.[35] Thus, accepting Eratosthenes' measurement of 252,000 stades for the circumference of the earth, the geographer ought not to include the equatorial zone, since that in Strabo's view is uninhabitable.[36] Instead he should start his analysis with the Cinnamon Country (near the mouth of the Red Sea[37]), about 8800 stades north of the equator, in the South, and with Ireland in the North.[38] He categorizes regions from south to north according to greatest length of day in equinoctial hours. This list, starting at Meroe with thirteen hours and ending at an area north of the Sea of Azov with seventeen hours, is similar to that of the elder Pliny given in Appendix II. In the extreme north, Strabo denied the existence of a Thule island. To him the most northerly inhabited area was Ierne (Ireland), itself 'only wretchedly inhabitable because of the cold[!], to such an extent that regions beyond it are regarded as uninhabitable'. Likewise, if one were to go not more than 4000 stades (500 Roman miles) north from the centre of Britain, one would find an area near Ireland, which like the latter would be barely inhabitable.[39]

Strabo's idea of the shape of the inhabited world is defined as follows:[40] 'Let it be taken as hypothesis that the earth together with the sea is spherical. . . , though not as complete a sphere as if turned on a lathe. . . . Let the sphere be thought of as having five zones. Let the equator be conceived as a circle on it, and let a second circle be conceived parallel to it, delimiting the frigid zone in the northern hemisphere, and through the poles a circle cutting these at right angles. Then, since the northern hemisphere contains two-fourths of the earth. . . , in each of these fourths a quadrilateral is delimited. . . . In one of these two quadrilaterals. . . we say that an inhabited world is settled, surrounded by sea and like an island.' He goes on to suggest that the quadrilateral in which the Atlantic lies resembles in shape half the surface of a spinning-wheel, and that the oikumene (inhabited world) resembles a *chlamys*, Greek mantle. This suggests that the eastern and western extremities of the oikumene were thought of as tapering and convex. He estimated the length of the oikumene as 70,000 stades and its width as less than 30,000.

As the ideal method of mapping the world, he writes in far more cartographic terms than before, 'We have now inscribed on a spherical surface the area in which we say the inhabited world is settled; and anyone most closely modelling reality by means of man-made representations should make a sphere of the earth, as Crates did, mark off the quadrilateral on it, and inside this should place his map of the *geographia*, i.e. of the inhabited world. But one needs a large globe, so that the section mentioned, being only a fraction of it, may clearly show the appropriate parts of the oikumene, which will present a recognizable shape to users. If one can construct such a globe, it should be not less than 10 ft in diameter. If one cannot make it as big or not much smaller, one should construct a map of the oikumene on a plane surface at least 7 ft long. For it will make little difference if instead of the circles, viz. parallels[41] and meridians, we draw straight lines between which to place the *klimata* with the winds and the other differences, and the positions of parts of the earth relative to each other and to celestial phenomena.'[42] He goes on to say there is little point in making the meridians converge slightly in such a map, so was it rectangular, a forerunner of something like Mercator's projection? The point of making it 7 ft long is that it is intended to represent an oikumene with a maximum length of 70,000 stades, so that we have an attempted scale of 1 ft to 10,000 stades, and the height will be nearly 3 ft. Taking 8 stades to a Roman mile, the scale becomes 1:6,250,000.

JUBA AND MELA

Of these two authors it would have been more interesting if the first rather than the second had survived. Juba II (*c.* 48 BC–*c.* AD 23) received from Augustus in 25 BC the kingdom of Mauretania, shortly afterwards marrying Cleopatra Selene, daughter of Antony and Cleopatra. He was a scholarly man, introducing Graeco-Roman culture and writing many works in Greek. Those of geographical interest were on Libya, Arabia and Assyria. Juba's theory of the source of the Nile[43] had an effect on the mapping of Africa. According to him, it rose on a mountain in lower Mauretania not far from the Atlantic, formed a stagnant lake called Nilides, then went underground to another lake in Mauretania Caesariensis, and thereafter again underground. Juba also provided Pliny with information about the Canaries:[44] distances, natural features, traces of habitation, flora and fauna (the name Canaria came from the Grand Canary's huge dogs, two of which were presented to Juba).

A contrast with Strabo's work is provided by the very simple and popular *Chorographia* of Pomponius Mela (*fl.* AD 37–42), who was born in southern Spain.[45] The title *chorographia* means regional geography, but Mela's work covers the whole world region by region. It was written in Latin in three books under Gaius or Claudius; there is no evidence that it contained maps.

Mela's world is divided east and west into what he calls two hemispheres. This is not a scientific definition, but a rough division of the known world approximating to Asia on the east, Europe and Africa on the west. From north to south he divided it into five zones, two cold, two temperate and one hot. This is a different approach from that of Strabo, who chose to ignore, as virtually uninhabitable, everything south of the latitude of southern India. It corresponds, however, to the division in Eratosthenes' lost poem *Hermes*, paraphrased by Virgil,[46] which regards the equatorial zone as 'altogether burnt up' but says that Antipodes live in the southern temperate zone.

Mela was writing before the Roman invasion of Britain, and has only a very rudimentary idea of its geography. Thule in his work does not sound like Orkney or Shetland. He says it is opposite the Belcae, the name which he uses elsewhere as a synonym for Scythians. One may therefore wonder whether he is thinking of an island north of Russia, or whether it is really some part of Scandinavia. This latter, however, is treated not as part of the continent but as a very large island. The Baltic is to him the Codanus Gulf,[47] enormous and dotted with large and small islands. Later he adds: 'In the gulf which we have called Codanus the most important island is Codanovia,[48] still settled by Teutoni; this surpasses the others not only in fertility but also in size.'[49] Mela's concept of Africa is less developed than those of later authors such as the elder Pliny,[50] but he does summarize Hanno's *periplus* (p. 132). He came from near Gibraltar, yet he believed there were no inhabitants of the central part of western Mauretania. His contribution to the disputed topography of Tartessus (Tarshish in the Old Testament) is to tell us that some thought it was on the site of Carteia, near Algeciras.[51] This is the kind of information that could have been entered in notes accompanying a map rather than on a map itself.

THE ELDER PLINY

We are fortunate in possessing all thirty-seven books of the greatest

Latin encyclopaedia, the *Naturalis Historia* of the elder Pliny.[52] C. Plinius Secundus (AD 23/24–79) came from Como. From 47 to 57 he was a cavalry officer in Germany, after which he became an advocate and writer. He rose, under Vespasian, to be procurator of Gallia Narbonensis in 70, of Africa in 72, of Hispania Tarraconensis in 73, of Gallia Belgica in 75. In 76 he was recalled to Rome and given an important post at the emperor's court. When in 79 he was admiral of the fleet at Misenum, on the northern extremity of the Bay of Naples, he saw the eruption of Vesuvius and sailed across the bay to give help and make observations, but was overcome by the fumes and died.

The elder Pliny obviously worked hard at his encyclopaedia, constantly making notes from Greek and Latin authors even as he was travelling round on a litter. But does this mean that he failed to observe? Book I consists of a listing of his sources for each of the remaining thirty-six books. In his text he often paraphrases these, whether Greek or Latin, fairly closely, sometimes mentioning the author's name, more often not (it cannot be called plagiarism, since he has briefly mentioned all sources), though quite a number are able to be conjectured.

Geography is treated in Books III–VI, cartography not in general being allotted a separate section but mentioned incidentally as a sub-division of geography. An exception to this is the closing sections of Book VI. From *NH* vi.206 Pliny first gives the dimensions of seas, then those of the three continents, basing his figures on Polybius, Artemidorus and Agrippa. His sources for this short section, extending to vi.210, are thus one Greek historian who had travelled extensively and had written on geography, one Greek geographical writer, and either the map or the notes of Agrippa, none of the writers being at all recent. This section is followed by the final one of the book (vi. 211–20), which he introduces with the words 'To this we shall also add one most ingeniously subtle theory of Greek invention,[53] so that. . . we may learn what are the affinities of days and nights in each region and which regions have equal shadows and an equivalent curvature of the *mundus*' (see Appendix II). He then outlines seven 'parallels',[54] approximating rather to *klimata* than to latitudes, with the data given in Appendix II. In Pliny's list, however, neither the shadow lengths nor the longest days are spaced at regular intervals.

The allocations may be criticized both in known and in less

known areas. For example, in Parallel 2, alongside Jerusalem (actual latitude 31°47′ N.) and Beirut (33°52′ N.) we find Libybaeum, the modern Marsala in west Sicily, whose latitude is 37°48′ N. Since mid-Sicily is in parallel 3 and north Sicily in Parallel 4, this has the absurd result of distributing Sicily among three 'parallels'. Again, we find part of Gallia Narbonensis and Nova Carthago (Cartagena) both in Parallel 4, though their latitudes are very different. Yet these and other such examples do not point to anything like climatic zones, since the maximum daylight hours are given for each. As in most ancient cartography, there is no appreciation that India extends as far south as 8°04′ N., corresponding in latitude to southern Sudan. In fact, surprisingly, in view of the work done by Alexandrian scholars, Parallel 1 in Pliny goes for Africa no further south than Lower Egypt,[55] and even includes Carthage (actual latitude 36°54′ N.). Pliny's lack of uniformity over gnomon lengths (Appendix II) and the fact that his shadow lengths fail to rise in geometrical progression are due, no doubt, to the fact that he selected from various Greek and Roman writers. He is not likely to have carefully sifted the whole available material based on or prepared for a map or maps; though some of his sources certainly incorporated these.

Books III–IV of Pliny's *Natural History*[56] deal with Europe, whose placing first in the geographical books he explains as follows: 'First then, Europe, nursling of the people that conquered all nations and by far the most beautiful of lands, has rightly been made by most writers not a third of the earth but a half, dividing the whole earth into two parts from the Tanais (Don) to the Straits of Cadiz.'[57] Within this section, Book III deals successively with southern Spain and coastal districts of Spain; southern Gaul; Italy, over which he becomes quite ecstatic: 'chosen by the will of the gods to make heaven itself more splendid, to unite scattered empires, . . . in brief, to become the single fatherland of all nations throughout the world.'[58] To conclude, he describes islands near Spain and Italy; the Alps; Dalmatia, Istria and the Danube provinces.

Book IV covers Greece, Macedonia, Thrace, and Greek islands except those near Asia Minor; the Black Sea and also northern Europe. Pliny's list of islands in the Baltic agrees closely with that of Pomponius Mela. When he comes on to Britain, we can see how lacking in up-to-date information his account is.[59] There is no mention of towns or tribes. The only reference to Roman occupation and settlement is in a subordinate clause oddly sandwiched between Pytheas' and Isidorus' measurement of its coast as 4875 miles and Agrippa's length and breadth of 800 and 300 miles. This runs: 'About thirty years ago Roman arms extended its knowledge no further than the area of the Calidonian forest.' Such a statement reveals his lack of serious study, and perhaps indicates that he did not make or even have access to maps of the area. The reference should be to the Claudian invasion, but there is no reason to think that this reached Scotland, where the Calidonian (or Caledonian) forest must have been. After Britain, Pliny describes Gaul apart from the Narbonese province, then the parts of the Iberian peninsula not previously described.

Book V opens with Pliny's description of Africa. In his reference to Hanno's voyage (v.8), we can see that his research was not careful, since he writes, 'There used also to be notes of the Carthaginian leader Hanno'. As mentioned below (p. 132), the Greek translation of Hanno's voyage is still extant. Either this or some Greek or Roman translation was in existence in Pliny's time, since he speaks of many Greek and Roman writers following Hanno, and he himself actually paraphrases one such in his account of an island off the west African coast (p. 48).[60] On more recent exploration of the west coast of Africa he quotes Polybius as having been supplied by Scipio Aemilianus with a fleet to sail along the coast. But the only item of information which he goes on to abstract is the one on *saltus* already quoted (p. 60). After this he goes on to the distances reported by Agrippa (pp. 47–8). Pliny's own interest lies mainly in the penetration of inner Mauretania by Romans during his lifetime.[61] Under Claudius, Roman forces and even allegedly civilians penetrated as far as the Atlas range.[62] Suetonius Paulinus, consul AD 66, was the first Roman commander to cross the Atlas range as far as the R. Ger;[63] Pliny summarizes the general's report. To what extent such reports were accompanied by maps or itineraries is disputed.[64] Although for the country of the Garamantes Pliny is able to record the names of places symbolically represented at the triumphal procession of L. Cornelius Balbus, 19 BC, for much of the interior of Africa he has recourse to tales of monstrous tribes: the Blemmyes or Blemmyae, who have no heads, the Himantopodes, who have thong-like feet and crawl instead of walking, and so on. For the source of the Nile he follows Juba (p. 65), with effects which lasted even up to the Age of Discovery.

After Africa, Pliny goes on to Palestine and Syria, followed by Asia Minor (apart from areas near the Black Sea) and islands off its coast including Cyprus, Rhodes and the nearest Aegean islands, followed by the whole area round Byzantium. This section is more in tune with Classical knowledge and better documented, towns often being quoted by more than one name if they have changed names.[65]

Book VI opens with a description of the Black Sea and areas of Asia near it, then on to Armenia and the Black Sea islands. Further east he is again in a poorly explored area and lets his imagination run wild. But he insists that the so-called Caspian Gates, where actual iron gates had been placed to stop migrations, were really the Caucasian Gates, thought to be where the main road from the Black Sea to Tbilisi now runs.

After describing tribes north of this area, he says of the Caspian: 'For the sea breaks in from the Scythian Ocean to the further parts of Asia, variously named by the local inhabitants, but best known by two names, Caspian and Hyrcanian. . . It breaks in by a narrow and very long mouth'.[66] But later (vi.51) he has to explain how Alexander the Great and Pompey's army drank sweet water from the Caspian: 'No doubt the salt was beaten by the quantity of river water coming in'.

Much of the information on the Persian Empire and India comes from the accounts of Alexander's geographers and ethnographers, though it is not always very accurate. Thus if Alexander averaged over 600 stades (75 Roman miles) a day sailing down the Indus and took a few days more than five months,[67] the distance travelled downstream would have been more than 11,475 Roman miles. Nevertheless it provides information about the interior which is valuable in the absence of his original sources. But it also includes travellers' tales, such as the two-hundred-year life-span of the Pandae (in the anthropological section, vii.28). For Sri Lanka[68] he records information both from Hellenistic sources and from a freedman of Annius Plocamus, who in Claudius' principate collected taxes from the Red Sea area. The freedman, carried to Taprobane (Sri Lanka) by adverse winds, spent six months there, learnt the language, and met the king; his visit resulted in four envoys being sent to Rome. Unfortunately, if he learnt the measurements of the island, they have not come down to us, and those of Eratosthenes quoted by Pliny, 7000 × 5000 stades (875 × 625 Roman miles as counted by Romans), are much too high. That measurement clearly accounts for the large island which appears on maps well into the sixteenth century. The route from Egypt to India,[69] important for its trade in Pliny's time, is described in detail. At midsummer one travels up the Nile to Coptus (Keft), then takes a camel-train to Berenice on the Red Sea, usually sailing from there via the Arabian port of Ocelis (perhaps Sheikh Said, near Bab-el-Mandeb). From there, if the west wind blows, it is forty days' sail to Muziris (Cranganore), the first trading post in India; the return journey is made in December or before 13 January.

Next Pliny tackles Mesopotamia,[70] where, after attributing the shift in distance of Charax from the sea (p. 50) to river deposit, he says that despite his preference for local authors he will not follow Dionysius, the most recent writer on regional geography, who was born there.[71] Instead he turns for that area and Arabia to reports by the Roman prefect of Egypt, M. Aelius Gallus, who in 25–24 BC led an expedition to Arabia Felix, and by King Juba, who sent a description of the area to Gaius Caesar, grandson[72] of Augustus. Pliny considers the Arabs the richest races in the world, owing to a mixture of brigandage and trading in produce of the sea or of forests (perhaps he is thinking of spices, in which they traded at second hand), buying nothing in return.[73] After this he writes of Upper Egypt and Ethiopia, and of islands off Africa.[74] On one group of these, the Gorgades off the west coast, he refers to Hanno's story of finding women covered all over with hairs, and taking the skins of two of these to Carthage. For the Canaries, which evidently only had abandoned buildings, he turns to Juba (p. 65). Book VII is concerned with geographical and other records which are mainly anthropological.

Thus, although, as has been seen in connection with his use of Agrippa's map and commentary (Chapter III), the elder Pliny consulted maps, he seems neither to have tried to compose them nor to have sifted his readings to analyse them as the scientist he tried to be. Instead, he seems, like many journalists, to have been obsessed with the unusual and the horrific, and to have exploited fears of the unknown. Yet the work is important to an understanding of cartography, for it was long after the Age of Discovery that strange peoples and animals disappeared from maps.

CHAPTER V

PTOLEMY AND HIS PREDECESSOR MARINUS

The period from about AD 50 to 150 was one in which Greek theory and Roman practice combined to produce great advances in mathematical cartography. These advances were particularly associated with scientists centred on the coast of the eastern Mediterranean, especially Alexandria, which had been a notable centre of applied mathematics ever since the foundation of the Alexandrian Library. In the earlier part of this period Hero(n) of Alexandria devised an astronomical method, not fully developed, of establishing the exact distance in a straight line between Alexandria and Rome.[1] As the Roman Empire now extended to Britain (from AD 43) in the west, Dacia (from AD 101) in the north and parts of Mesopotamia (AD 114–17) in the east, and as the construction of Roman roads accelerated contact with overseas colonies, new knowledge and information of distant places was flooding in. Not all Greek writers, however, could read Latin well, and some did not visit much of the western Roman world. This language and cultural problem could account for some of what to us are surprising gaps and slowness in the incorporation of new data.

MARINUS

Marinus of Tyre is only known to us through criticisms, mainly hostile, in Ptolemy's *Geography*. Marinus was working about AD 100 or 110.[2] Although he lived in Phoenicia, he wrote in Greek; his name looks Roman (= 'marine'), but may in fact not be derived from Latin, but from Greek ('a kind of sea-fish') or Semitic. His publication is called by Ptolemy *Correction of the World Map*.[3] Later Ptolemy says that Marinus never actually drew a map to illustrate his latitudes and longitudes 'even' (if this is correct: manuscript readings and translations vary) in his last edition.[4] Ptolemy also states that those who have composed maps 'according to Marinus' have too hastily worked on notes instead of following the latest systematic treatise (*syntaxis*, the word he uses of his own astronomical treatise). Perhaps Ptolemy did not consider any of these worth acquiring: he complains of the difficulty of working from Marinus' text. There seems no doubt that mapping attributed to him was circulating in much later times, since it is actually mentioned in Arabic geographers (p. 177). Modern reconstructions assume that, since Ptolemy criticizes Marinus for what he considers major errors, areas of the world for which he does not criticize him had a similar appearance to that which can be reconstructed from Ptolemy's co-ordinates. This seems a reasonable assumption in view of the praise which he accords Marinus before proceeding to embark on hostile criticism.

He first refers to Marinus immediately after his statement that it is important to collect all the latest data available.[5] He continues (i.6): 'Marinus of Tyre seems to have been the most recent of our students of *geographia* (= map-making) and to have applied himself to this subject with the greatest enthusiasm. . . . If we could see that his latest composition lacked nothing, we should even have been happy to complete our description of the known world from these notes of his alone, without researching any further. But as on certain points he himself seems to have composed without reliable comprehension, and as in embarking on his map he has in many places not devoted enough thought either to convenience or to symmetry, we were naturally induced to contribute to his work what seemed necessary to make it more logical and useful. We shall do this as simply as possible. . . , first from a study of scientific exploration, as a result of which Marinus thinks that the length of the known world should be extended towards the East and its width towards the South.'

Ptolemy first proceeds to explain the concept of length and width, both in terrestrial and celestial cartography, then examines the width and length, in that order, of Marinus' map. On its width,[6] he says that Marinus placed the island of Thule right up to the parallel which delimits the most northerly bound of the known earth, namely about 63°N. of the equator, or 31,500 stades (= 3937½ Roman miles), if one reckoned 500 stades to a degree. Marinus placed the land of the Ethiopians called Agisymba and the

Prasum promontory right up to the parallel delimiting the most southerly bound of the known earth, on the winter tropic; and according to Marinus the width of the known world was 87° or 43,500 stades (= 5437½ Roman miles). Ptolemy then quotes verbatim from Marinus an astronomical justification of the placing of Ocelis: 'In the torrid zone the whole zodiac is above one . . . ; only the Lesser Bear begins to appear in entirety in areas 500 stades north of Ocelis. The latitude of Ocelis is 11°24′ N. Now Hipparchus says that the southernmost star of the Lesser Bear, the last in its tail, is 12°24′ distant from the pole . . .'. This led Marinus on to the stars visible to those travelling either by sea to India or east Africa or up the Nile.[7] Marinus calculated,[8] from the number of days said to have been spent on land journeys, that Agisymba was 24,680 stades (= 3085 Roman miles) further south than the equator, and from the number of days on sea journeys that the Prasum promontory was 27,800 stades (= 3475 Roman miles) further south than Ptolemais Troglodytica, though Ptolemy criticizes such a measurement as too short. The land measurement was based by Marinus on expeditions of unknown dates by two Romans, Septimius Flaccus and Julius Maternus.[9] But 24,680 stades south of the equator would give over 49°; this, if added to the approximate figure of 63° north of the equator, would give over 112°, which he thought rather long.

For the length, or west–east measurement, Marinus reckoned a fifteen-hour (225°) interval,[10] which Ptolemy also considered too long. Marinus calculated this likewise by land and sea. A Macedonian trader, Titianus Maes, is said to have measured the land distance to China, not by going there himself but by sending men out there. By Ptolemy's reckoning it is only 177°15′ from the Canaries to Sera, supposed capital of China. For the sea journey Marinus quoted a navigator, presumably Greek, called Alexander,[11] who sailed first to the Golden Chersonese (Malay Peninsula) and from there to Zabae and then southwards to Cattigara. Since Marinus did not quote distances beyond the Golden Chersonese, only 'several days' from Zabae to Cattigara, the measurement is very inexact and unscientific.

Ptolemy next criticizes Marinus for his location of certain places on the Mediterranean 'opposite' certain others. For example, Marinus stated that Pachynus in Sicily was opposite Lepcis Magna, and Himera in Sicily opposite Thenae (12 km S. of Sfax, Tunisia), whereas from Pachynus to Himera is only about 400 stades and from Lepcis to Thenae about 1500. There are also comparisons related to latitude belts, for example: 'Whereas he said Noviomagus [Chichester] was fifty-nine miles south of London in Britain, in the *Climata* he shows it [i.e. on the map] as being north of London.'[12] Rather than imagining that Marinus was confusing two places in Britain, should we take it that he had a list of select places in latitude belts, not correlated with the earlier information? The figure of fifty-nine miles applies not to the road distance between London and Chichester[13] but to the distance as the crow flies from the south bank of the Thames, near London: the difference in latitude expressed in Roman miles is about forty-eight.

Finally, Ptolemy objects that much of Marinus' work is too difficult and diffuse to follow without a map to hand. For example, the student may have to turn to one passage for the latitude of a particular place and to another for its longitude. By the rectilinear projection adopted in Marinus' map, only at the parallel of Rhodes do distances north–south and east–west occupy the same measurement on the map. This seems to have induced Ptolemy to look for more realistic projections of a map of the known world on a flat surface. To what extent topographical aspects of Marinus' map were detailed and of the same nature as Ptolemy's we cannot tell. But there were undoubtedly features of it which Ptolemy could not criticize and is likely to have borrowed.

PTOLEMY

Claudius Ptolemaeus, known to us as Ptolemy, lived from about AD 90 to about 168. Of his names, the first is Roman, the second that of the Macedonian dynasty of Egypt; his language, like theirs, was Greek. He is said to have been born in Upper Egypt, and he lived most of his life in Alexandria, where from 127 to 141 he made astronomical observations. Geography came later, and was also approached from a mathematical angle.

His *Mathematical Syntaxis*,[14] composed between AD 141 and 147, is the same work as was known from the Middle Ages as the *Almagest*, a name derived from Arabic *al* + Greek *megiste*, 'the greatest (compilation)'; it indicates the route by which Ptolemy first became known in the West. In this treatise he gives facts and figures for drawing a map of the stars. His work, relying on observations made up to AD 141, incorporates a catalogue of all known stars (Books VII–VIII), based on Hipparchus' catalogue, with ecliptic co-ordinates. As he planned this to be a permanent list, wherever the traditional names seemed inappropriate (e.g. where one could improve on the part of an animal's body incorporated in a name) he altered them. He recommended as the ideal celestial map a dark globe, with two great circles at right angles on it, one of these representing the zodiac. With the help of two metal semicircles, one rotating with the sphere and one independently, 1022 stars were to be placed at the intersection of the correct latitude and longitude on the globe. This approach represented an advance on earlier celestial mapping, which had concentrated on constellations rather than individual stars. It was also an advance to use ecliptic rather than equatorial co-ordinates.

Although it is mainly concerned with celestial mapping, the *Mathematical Syntaxis* does touch on terrestrial mapping too. He takes it that the earth is spherical in a geocentric universe.[15] 'The earth', he writes, 'may be divided into four by the equator and a meridian; the inhabited world is approximately contained in one of the two northern quarters'. Within that framework, latitudes of select places were to be plotted by length of daylight on the longest day; this was a concept taken from Hipparchus. The most southerly inhabited place was Taprobane (Sri Lanka) with 12¼ hours; the most northerly Thule, with 20 hours. Whereas he started with the idea of thirty-two parallels, he reduces these to eleven, and his final plan was for seven, a number which persisted throughout the Middle Ages.[16] These, based on daylight hours ranging from thirteen to sixteen hours, did not cover the whole oikumene but the populous parts as then known, with lines going respectively through Meroe, Syene (Aswan), Lower Egypt, Rhodes, the Hellespont, the centre of the Black Sea and the R. Borysthenes (Dnieper). He then comments that he will deal in a later work with the details of 'this fascinating subject' relying on the records and surveys of others.[17] His project was fulfilled about twenty years later in his geographical work, where, however, the point of origin for longitude is not, as in the earlier treatise, Alexandria but the Canary Islands.

The title of the later work is *Geographike Hyphegesis*, 'Manual of Geography', but it is normally known today as *Geography*.

Technically, it is not Ptolemy's only work on the subject, as there is also a brief *Canon of Significant Places*, which often differs somewhat both from Books II–VII (co-ordinates) and from Book VIII of the *Geography*; these two themselves do not necessarily agree.

In his opening words of the *Geographike Hyphegesis*, '*geographia* is an imitation (mimesis), by drawing, of the whole known world and its features'.[18] As Letronne observed: 'Ptolémée prend le mot géographie dans le sens graphique et non descriptif',[19] perhaps something akin to graphicacy as defined by Balchin. Ptolemy outlines the features mentioned as including the larger cities or towns,[20] mountain ranges and the chief rivers. Here some sort of symbolic representation seems to be implied. Geography, as he envisages it, is complemented by chorography, which studies detailed topography, such as harbours, villages or demes, and the details of river flow. As to records and surveys, he quotes Marinus of Tyre as having discovered much information, and despite criticisms obviously treats him as the main source. But he also had Greek theoretical sources to consult, and Roman topographical works, some of which he misunderstood; thus Alba Fucens becomes Alpha Bucens, and an imaginary name in Germany, Siatutanda, is a misunderstanding of a passage of Tacitus[21] explaining the movements of a German tribe as designed *ad sua tutanda*, 'to protect their possessions'.

After criticizing Marinus' orthogonal projection, Ptolemy continues: 'One would therefore be right in keeping the meridian lines straight, but the parallels as areas of concentric circles. . .'.[22] In contrast to his attitude to celestial cartography, he recommended mapping the world on a flat surface, but with conic projection. In his first system, to which the above statement refers, the meridians were drawn as straight lines converging on a point representing the north pole, and all the parallels as arcs with this point as centre. He writes: 'The Rhodes parallel must be included, because many proved distances have been recorded on it'. Since it was a map of the known world only, the north pole would be well outside the mapped surface. He needed, however, to decide how far the known world could be said to extend before working out the proportions of the map.

Marinus, he felt, had made it too big. From north to south, although Ptolemy accepted Thule, at 63°N., as the northernmost point known, he did not go as far south as the Tropic of Capricorn. Instead, he gave as the southern limit Agisymba and Prasum promontory, at 16°25′ S. This made the north–south distance about 40,000 stades. From east to west he decided on a longitude of 12 instead of 15 hours. He reckoned a total of 177°15′ from the Canaries to Sera, supposed capital of China, and made it up to 180° by extending it as far as Cattigara (Kattigara if we accept the Greek spelling). This gave 72,000 stades as the length of the oikumene on the parallel of Rhodes, where one degree of longitude was reckoned as 400 stades. Hence he had to provide for a proportion of 5:9. Graeco-Roman maps hovered round the 1:2 proportion for the known world; north–south occupied the smaller distance. Ptolemy's north, from his comments on projection, appears to have been at the top of the map.

Ptolemy recommended the drawing of 36 + 1 meridians at twenty minutes of time, or 5°, on the equator. His parallels were planned thus: one at half-hour intervals of daylight length south of the equator, one each at quarter-of-an hour intervals of daylight length from the equator to 45°N., and one each at half-hour intervals of daylight length from there to 63°N. The reason for what appear as unequal measurements on reconstructed maps is that more exact observations had been taken in the better known areas. He held that places about which there was uncertain information should be reckoned in relation to more definitely fixed places.[23]

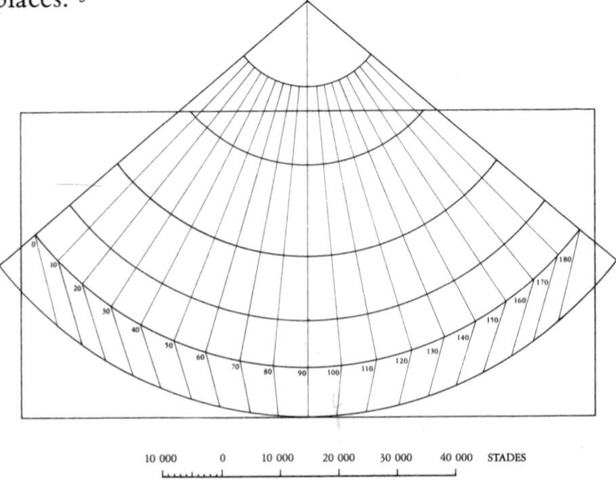

Fig. 9. Ptolemy's first projection.

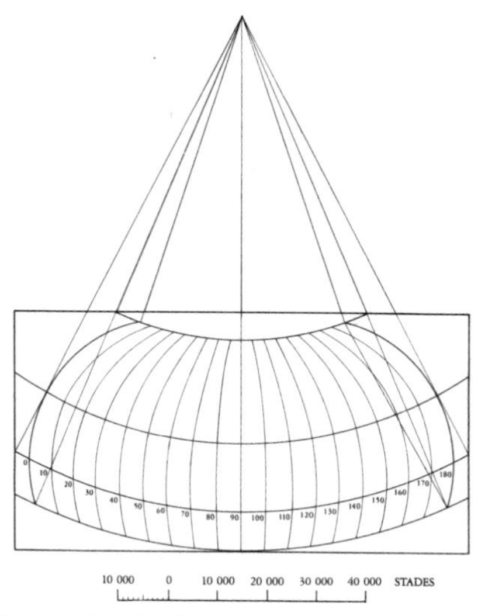

Fig. 10. Ptolemy's second projection.

The chief difficulty about his first projection is that when it reached the equator it suddenly made an abrupt angle. A conic projection based throughout the oikumene on the north pole would give an enormously exaggerated scale south of the equator, so that he had to modify it by adding south polar conic meridians, resulting in an unnatural bend in coastlines. Hence he proposed a second projection, which would 'allow maps to be drawn closer to reality'.[24] The aim of this was to give the meridian lines 'the shape which they appear to have on a globe when an observer's visual axis passes through the intersection of the central meridian with the central parallel, and also through the centre of the globe.'[25] Ptolemy took as his central parallel for this purpose the latitude of Syene (Aswan), 23°40′ N. The eighteen meridians on each side of the central one were made to curve in due proportions, so that distances north–south and east–west could be better equalized.

With this system it is particularly difficult to draw places in correctly. A third projection suggested later in the work[26] did not improve this aspect. So Ptolemy recommended the first projection of the three for a map of the known world, and orthogonal grids for the regions. According to its mean latitude, the grid for each regional map was to be constructed with proportional distance between east–west and north–south divisions,[27] e.g.

Regional Map	*Proportion*	*Percentage equivalent*
British Isles	11:20	55
Great Germany	3:5	60
Gaul	2:3	$66\frac{2}{3}$
Spain, Italy	3:4	75

Thus the different shapes and sizes of regional maps are dependent not only on the shapes of regions but on the grid structure.

The dispute which long raged whether Ptolemy drew maps for his *Manual of Geography* has largely been settled.[28] In Books I and II he uses the future tense, 'we shall draw', and many scholars long maintained that he was merely guiding others, who may have had Marinus' map, to compose their own maps if they wished. If one were to follow the Renaissance translation of *Geog.* viii.2.1, one could come to the same conclusion for that passage. But in fact the Greek means 'we have had maps made'. The maps there listed are ten of Europe, four of Africa and twelve of Asia. Whereas in earlier Books the places are listed by degrees and minutes of longitude and latitude, here Ptolemy says they have been entered by hours east or west of Alexandria and by maximum daylight hours. If we follow Polaschek, we shall deduce that Book VIII was never revised and that Ptolemy commissioned regional maps for his first edition only.

The most specific reference[29] to a Ptolemy map which is likely to antedate our manuscripts in origin is the statement of Agathodaimon: 'I, Agathos Daimon (= "good spirit", usually contracted to Agathodaimon), a technician of Alexandria, drew a map from the *Geography* of Ptolemy.' This statement is inserted in eight manuscripts at the end of Book VIII.[30] Since he was a Greek technician working in Alexandria, he can hardly have lived later than the sixth century AD and there is nothing to disprove a very much earlier date, e.g. soon after Ptolemy's death.[31] The map which he drew was presumably the world map. It is true that the maps of Codex Vaticanus Urbinas graecus 82 (late thirteenth century) which we possess are very close to the text. But this is surely no proof, as J. Fischer claimed,[32] that they are descended from maps compiled by Ptolemy himself; and if they were descended from unofficial but very early Ptolemaic maps, one might expect less accuracy. If we compare the maps in manuscripts of the Corpus Agrimensorum, we find that repeated copying, which was the regular practice, led to serious corruption. Nevertheless, anyone could at any time have made world or regional maps from the instructions in the *Geography*.

After the introductory Book I, Books II–VI and the first half of Book VII are devoted to the regions of the world, starting with Ireland in the far West and ending with the Far East. Each town, river mouth etc. is given a longitude east of the Canaries and a latitude. These are expressed in Milesian numeration, using α′, β′ etc. (with addition of digamma for 6) to ι′ for 1–10, then ϰ′-π′ for the tens from 20 to 80, koppa for 90, ϱ′ for 100; higher numbers do not occur in the co-ordinates. This numeration is used only for degrees: instead of minutes we have fractions of degrees, e.g. $5' = \frac{1}{12}^{\circ} = \iota\beta''$.[33]

The world map which results from Ptolemy's text may not have been very different in general effect from that of Marinus. But if so, then both differed in one concept, as far as we know, from Greek world maps ever since Anaximander. The orthodox idea was that the known world was completely surrounded by a continuous ocean.[34] Ptolemy, however, trusted Marinus' account of the navigator Alexander, who said that Cattigara was appreciably to the south of previous landing places. As a result, he considered that the Indian Ocean was enclosed on each side, with an imaginary southern land somehow linking the Far East to east Africa.[35] Where Cattigara actually was is uncertain, but one theory is that it was Hanoi, which may have been called Kiau-chi, so that Cattigara = 'Kiau-chi town'.[36]

It is clear that coastlines were important for the framework of Ptolemy's *Geography*. Rivers are given co-ordinates only at their mouths, and promontories and bays too are listed with co-ordinates. Modern reconstructions based on these features present an angular appearance as opposed to extant maps of the late Byzantine and Renaissance periods (Chapter XI). This interest in coastlines may explain a greater discrepancy between groups of manuscripts in their co-ordinate readings for coastal than for inland features. Research by E. Polaschek makes it seem likely that different groups of manuscripts represent successive recensions of the co-ordinates in antiquity, at least the first of which may be attributed to Ptolemy himself.[37] The most striking example of this effect is in southern Italy, where among other changes the 'heel' of Italy is reorientated. Although in some cases we can dismiss variants as corruptions in the Milesian numeration, in others this seems less likely. A section of the British coastline where there may be indication of a revision is discussed on p. 83.

Much has been written on the details of Ptolemy's regional geography. The following is only a brief outline of points referring to a small selection of regions.

(a) *The British Isles.* The co-ordinates for Ireland, the first of the regions in Book II, indicate that Ptolemy places the island about 6° too far north. This puts it between 57° and 61°30′ N., occupying much the same latitudes as north Britain north of the Mersey. The position of Ireland may have contributed to the conspicuous deformation of Scotland, for the Mull of Galloway's co-ordinates, 21°E. of the Canaries, 61°40′ N.,[38] make it the northernmost promontory. All land known by us to be north of it, as far as Tarvedum or Orcas promontory (Dunnet Head, more correctly Tarvedunum), is given co-ordinates which place it to the east: Dunnet Head is marked as being 30°20′ E. of the Canaries, 60°15′ N. Much has been written on the origins of and reason for this deformation.[39] The measurements of Britain recorded by Diodorus Siculus,[40] 7500 stades for the south coast, 20,000 for the

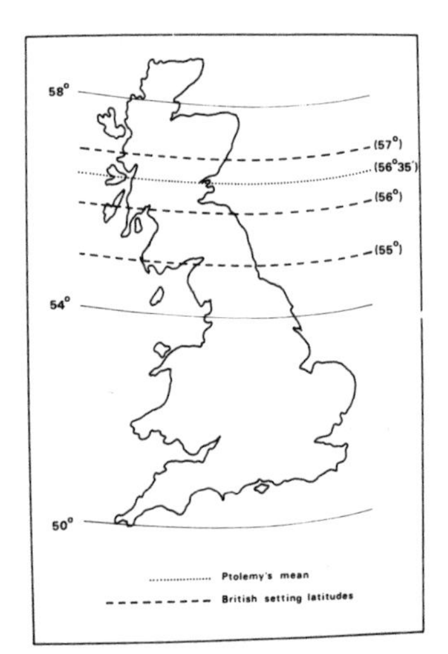

Fig. 11. Latitudes for Britain in Ptolemy and on Roman sundials. Mean latitudes for provinces are given on the reverse of some portable sundials of later date than Ptolemy, and three of these approximate to his incorrect mean of 56° 35′. See also pl. 6.

west and 15,000 for the east, imply a large triangle (even if coastal indentations are included) with apex well to the east. They may have been taken from Eratosthenes' map, which is likely to have influenced Marinus and through him Ptolemy. Although Agricola in the territory of the Boresti (Moray Coast) had ordered the commander of his fleet to circumnavigate Britain in AD 84, the latter may not have recorded orientations. In areas where there were Roman roads, distances would be known, but not necessarily anything more. Ptolemy's second world map projection, whose meridian lines have a greater curve as they approach the east and west extremities, may also have affected the orientation, especially as Scotland would be on the north-west margin. He knew that the length of Britain was the equivalent of about 800 km. He probably also had some orientation of the occupied areas as far north as about the R. Wear. If he continued on the same line, he would have made the mainland of Britain extend too far north to allow Thule, which for him was Shetland, to be north of it and yet to be within Marinus' latitude, which he had accepted. If he orientated the southern half of Britain roughly north–south instead of NE–SW as Eratosthenes may have, he may possibly have thought that a correction needed to be made for the northern half.

Ptolemy's information on Britain, consisting chiefly of co-ordinates, is in Book II, Chapter 3 of the *Geography*. It may be consulted in Müller's edition,[41] with Latin translation, or in Rivet and Smith;[42] and a reconstruction from the co-ordinates is given in the Ordnance Survey map.[43] All of these give only one set of co-ordinates for each place, based on probable readings. But it will be seen from Müller's apparatus criticus that many variants exist, and these are listed in Appendix IV. The variants most likely to represent a revision are on the promontory of the Novantae (Mull of Galloway). Most manuscripts give the latitude of the R. Abravannos, thought by Rivet and Smith to be the Water of Luce, as 61°, but one group has it as 60° 15′; and there is also a slight variant between these groups of manuscripts for the nearby Iena estuary. In a number of cases, minutes of longitude or latitude are rounded off to an exact number of degrees. The variants contribute little to correcting major misplacements. Thus Vinovium or Vinovia (Binchester), which is in north-east England, is misplaced by Ptolemy, in either variant, as being near the north-west coast, south of Moricambe estuary.[44]

For southern Britain Ptolemy seems to have relied on reports from AD 43 to about 70. For northern Britain he turns mainly to evidence of Agricola's times, 78–84, probably drawn from Marinus, but for the legionary headquarters at Chester and York to later evidence, incorporating changes of *c.* AD 87 and 119 respectively. It is not surprising that places on Hadrian's Wall (built AD 122–6) are absent, but those on the earlier Stanegate (AD 80) are also missing. Exeter is labelled, in translation, 'Iska, 2nd legion Augusta, a *polis* of the Dumnonii'. The legionary information in this entry is either out-of-date or mistaken. It has only fairly recently been established by excavation that Legion II Augusta was at Exeter in the early stages. By the time Ptolemy was compiling his list, which includes York (Eburakon, Legion VI Victrix) and Chester (Deva, Legion XX Victrix), Legion II Augusta had moved to the other Isca (Caerleon); with this duplication of names it is not surprising if updating by a compiler at a distance was missed.

As towns were listed in relation to the relevant tribes, siting was made easier. The most curious mistake concerns Urolanion, i.e. Verulamium, which Ptolemy gives as about 95 Roman miles from London,[45] whereas its true distance by Roman roads is about 25. Such a mistake could have arisen from dictation if kappa epsilon (25) were misheard as koppa epsilon (95). Isca (Exeter) is shown as about 115 Roman miles instead of 170. Calleva, known from an inscription[46] to be Silchester, is wrongly oriented. Durovernum Cantiacorum (Canterbury) is called by Ptolemy Daruernon or Daruenon and located somewhere near Maidstone. What is perhaps surprising is the large number of places named in Scotland. Since three of these are Korda, Koria and Kuria (variant Koria), one might wonder if there was duplication; but Korda is described as a *polis* of the Selgovae, Koria of the Dumnonii, and Kuria of the (V)otadini, so that Richmond and Crawford were probably right in thinking of Coria as a name for a tribal meeting-place.[47] The suggestion of R. J. Wyatt that in several instances, e.g. Vindogara–Corda–Carbantorigum, three places in northern Britain have been plotted in a straight line is not entirely borne out by plotting on graph paper.[48]

The choice of place-names is erratic: Gloucester, Caerleon, Caerwent, Chelmsford, Manchester and Corbridge do not appear, while a number of unimportant places do. Among the latter group we may perhaps try to identify one in the west. Rivet and Smith write of Uxela: 'Unknown, but apparently an early Roman fort in Devon or Cornwall'.[49] Now further east Ptolemy has the place Iskhalis (or Iskalis) which, wherever it was, looks as if it was on the Somerset Axe.[50] We therefore need a more westerly river for Uxela, given also as an estuary by Ptolemy: and the Ordnance Survey *Map of Roman Britain* seems right in equating it with the R. Parrett, or perhaps equally well its tributary the R. Cary. If so, the settlement Uxela must have been at or near Bawdrip, to which a Roman road led from Lindinis (Ilchester). The fact that the co-ordinates for Uxela and for the Uxela estuary are rather different is due to Ptolemy's separate plotting of coastal and inland features. In Italy this applies even to lakes, so that L. Como looks as if it is rather distant from Como.

(b) *The Baltic*. In the absence of accurate information, Ptolemy gives references for the shore of Great Germany, as far as the R. Vistula, which are absolutely straight west–east on latitude 56°. Scandinavia emerges as three small and one large Scandia island.

(c) *Italy*. Here the coastline cannot be understood without reference to the Po valley. Although Ptolemy does not give co-ordinates for roads, he does often list the principal places on them. In the case of the Po valley, the line of the places on the Via Aemilia shows that he mistakenly thought of it as running basically east–west, whereas in fact its orientation is roughly ESE–WNW, and his plotted points when listed are about 30° out. This mistake may have arisen if Ptolemy consulted one or more centuriation maps of the Po valley and imagined they had due north at the top; cf. (d) below. The effect of such orientation of the Po valley is to make the Adriatic coast from Rimini to Ancona extend too far eastwards, as a result of which he realizes that for the southern section some correction is needed. The heel of Italy seems to show evidence of three successive recensions, of which it is thought that the first two are those of Ptolemy himself.[51] But in each of these it has too north–south an orientation. The co-ordinates for the shape of Sicily (not part of Italy in Roman times) are less accurate than one might have expected: from an island so densely settled by Greeks it seems strange that he had not more accurate information.

(d) *Central north Africa*. Whereas due regard is paid to the Gulf of Benghazi, the references for the coast from C. Bon (Tunisia) to Brachodes promontory (Ras Kapudia) give a roughly NW–SE or WNW–ESE alignment instead of roughly north–south. Since

Ptolemy lived in Alexandria, one might have expected him to know this stretch better. But (i) he may, as with the Po valley, have looked at a Roman centuriation map and have taken the approximately NE-facing *kardines* as facing north, thus rotating the orientation by about 45°; (ii) travellers from Egypt to Carthage would tend to go by sea but keep well out and clear of the Lesser Syrtis sandbanks.

(e) *The Red Sea and Persian Gulf.* References shape the Red Sea too broad towards the south. Co-ordinates for the Persian Gulf form a rectangle, with greater length east–west. Its northernmost point is given the same latitude as Alexandria, somewhat too far north; its southernmost point, owing to the shape, is much too far north.

(f) *India and Sri Lanka.* Ptolemy's figures for the Indian coastline give the angle between the east and west coasts as less acute than it is and much too far north, so that the effect must be too flattened a coast. The size of Taprobane (Sri Lanka), as in other Classical writers, is excessively large.

(g) *South-east Asia.* Mention has already been made of Cattigara, the furthest place east on Ptolemy's co-ordinates (p. 81). The sequence before this is as follows: first, east of the Ganges Gulf is the Golden Chersonese, which corresponds, though smaller, to the Malay Peninsula. Then, to the north-east, is the 'Great Gulf', which he associates with the Chinese,[52] and Cattigara is on the far side of this. Modern investigators have wondered why at this point there is a turn southwards. A writer in South America advances the theory that the 'Great Gulf' is really the Pacific and the area on its far side a part of South America corresponding to Peru.[53] Although such a theory should not be dismissed out of hand, it is open to serious objections: (i) there is no evidence that Europeans in antiquity reached South America; (ii) granted that Ptolemy was working on a length for the oikumene of only 180°, much shorter than Marinus', rather than greatly shorten any distance he would have changed its orientation; (iii) the centre of the 'Great Gulf' is associated with Sinae, i.e. Chinese; (iv) in any such investigation, a distinction should be made between text with co-ordinates, as Ptolemaic, on the one hand, and regional maps on the other hand, since these are likely to have originated later (see Chapter XI) and merely to have been based on the former evidence as interpreted perhaps in the Byzantine period.

CHAPTER VI

LAND SURVEYING

GREECE AND ITS COLONIES

The Greeks surveyed within a system of squares or rectangles for towns and of rectangles for rural areas. Substantial remains of their urban rectangles may be seen in the layout of the central district of modern Naples.[1] The inspiration for this type of survey came, according to ancient writers, from Hippodamus of Miletus.[2] It must have been he who in 479 BC laid out his native Miletus, a city famous for early science and extensive colonization, on a system of large and small squares. Hippodamus was also commissioned after the Persian Wars to re-plan the Piraeus on a grid pattern. In 443 BC he was among those who took part in the foundation of Athens' only colony, Thurii in south Italy. The beginnings of this colony were on a small scale, and its original streets are said to have totalled only four in one direction and three at right angles. It is possible that the Milesian tradition antedated Hippodamus, since already at the end of the sixth century BC, after a fire, Olbia, a Milesian colony on the Black Sea, was rebuilt on a grid pattern.

Modern investigations have been able to show evidence of much orthogonal planning from the Greek world. For example, Priene in Asia Minor and Olynthus in the Chalcidice had a large urban grid. Megalopolis in Arcadia, founded in 371 BC as a centre for greater urbanization and better defence, had buildings on both sides of the R. Helisson on the same orientation. Recent investigations by Professor H. Williams, of the Canadian Archaeological Institute, Athens, with a proton magnetometer and computer mapping, have shown evidence that the city of Stymphalos in northern Arcadia (now south Korinthia) was laid out in rectangular blocks measuring approximately 103 × 30 m, with streets approximately 6 m wide;[3] clearly the combined measurement of block and street was intended to have a length three times its width. Investigation by aerial photography of Greek colonies in south Italy, Sicily and the Black Sea has yielded similar results. Two systems of land division have been observed in the territories of the adjacent Greek colonies of Metapontum (Metaponto) and Heraclea (Policoro) in south Italy.[4] They are visible in belts extending 6 and 7 km respectively, and are of very long strips about 230–240 m wide, the easternmost having the same orientation as the streets of Metapontum. It is thus increasingly obvious that Greek surveyors or 'measurers of land'[5] were extremely practical and in Italy passed on their knowledge to the Etruscans and Romans. What we do not know is whether for their town and country surveys they constructed maps and plans. It is difficult to conceive how such accurate ground measurement would have been achieved without some form of 'blueprint'.

ROMAN SURVEYS AND ASSOCIATED MAPS

The Romans, from quite early times, mainly favoured a system of squares in which to draw up surveys. In town areas these squares were *insulae* (blocks) of very varying sizes.[6] In country areas they were *centuriae* ('centuries'), which were most commonly squares of 2400 × 2400 Roman feet. Surveyors were known as *mensores*, 'measurers', and land surveyors as *agrimensores*, 'land measurers'.[7] Owing to the existence, in part, of a collection of their textbooks, the Corpus Agrimensorum, containing treatises from the first century AD onwards, not only can we assess their importance as a professional body but we also have evidence of their use of maps and plans.

The standard square 'century' had an area which was theoretically of 100 *heredia*, 'heritable plots', but in practice of 200 *iugera* (*c.* 124.6 acres/50.4 ha.). For this purpose the *iugerum* represented a rectangle of 120 × 240 Roman feet. An *actus* (pl. *actus*), 120 Roman feet, constituted the basic unit of length. It literally meant 'a driving', the distance which oxen pulling a plough were driven before turning. The official work of the land surveyors was strongly linked to the foundation of colonies and the allocation of land to new settlers. To measure out the allotments they used a *groma*, a type of cross-staff consisting of a wooden and iron cross, with plumb-lines attached to its extremities.[8] The cross rested on an angle bracket, which swivelled on a pole whose iron tip could be fixed in the ground. Sighting seems to have been from

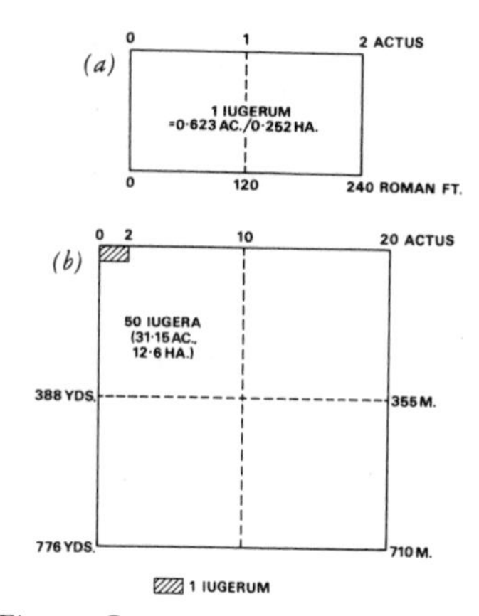

Fig. 12. Centuriation measurements: (a) 120 Roman feet = 1 *actus*, 2 square *actus* = 1 *iugerum*; (b) 20 × 20 *actus* (400 square *actus*) = 200 *iugera* = 1 *centuria* of standard size.

Fig. 13. Groma: reconstruction of the surveyor's cross, with swivelling bracket and plumb-lines, in use AD 79 and found in the workshop of Verus, Pompeii. Its metal parts are preserved at the National Museum of Antiquities, Naples.

one plummet to its diagonally opposite one (two different pairs of plumb-bobs were found), with the help of an assistant holding a ranging-pole. A straight line was laid out and measured with the *decempeda*, 10 Roman feet long. With the standard foot this would measure 2 m 95.7; with the *pes Drusianus*,[9] used in places in the north-west provinces, it would be 3 m 33 or 3 m 35. Metal ends of *decempedae* have been found in some places, though one may surmise that often the wooden shafts were not given end-pieces.

In addition to instruments proved to be associated with surveyors, Roman instruments survive which could have been useful for the drawing of maps and plans. These include folding foot-rules (1 Roman foot = 29.57 cm long), set-squares, dividers and compasses. In particular there are double bronze compasses, able to be tightened by a wedge, now in the British Museum,[10] which have the two ends in a proportion of 2:1, so as to facilitate drawing at half or double the existing scale.

The usual name for the system of land division was *limitatio*, but as 'limitation' in English has a different sense, it is usual in English writing to have recourse to the alternative word *centuriatio*, which is anglicized as 'centuriation'. *Limitatio* was named from *limites*: a *limes* was literally a balk separating two ploughed fields, but in centuriation schemes it normally became a road. The two main intersecting *limites* were called *cardo (kardo) maximus* and *decumanus maximus*, abbreviated to KM and DM, with other *kardines* and *decumani* paralleling them. Orientations varied, but Frontinus and Verrius Flaccus favoured *kardines* running north–south and *decumani* east–west. The *kardo maximus* and *decumanus maximus* were the first to be surveyed. Based on them, a map was drawn up by the surveyor, using the following abbreviations to form co-ordinates for the 'centuries':

DD	*dextra decumani*	to right of DM
SD	*sinistra decumani*	to left of DM
VK	*ultra kardinem*	beyond KM
CK	*citra kardinem*	near side of KM

Numerals were then added to the abbreviations, to indicate how far a particular 'century' was from the KM and DM, e.g. DD III VK XI. This type of co-ordinate is used in the stone cadasters of Orange (p. 108).

The remains of Roman centuriation are in places very conspicuous,[11] and traces have been recorded since 1833, when the Dane C. T. Falbe noticed 'centuries' of regular size near Carthage. The very large area in the Po valley is in the territory of colonies based on the Via Aemilia. Among numerous other areas in Italy which have remains of centuriation, mostly connected with

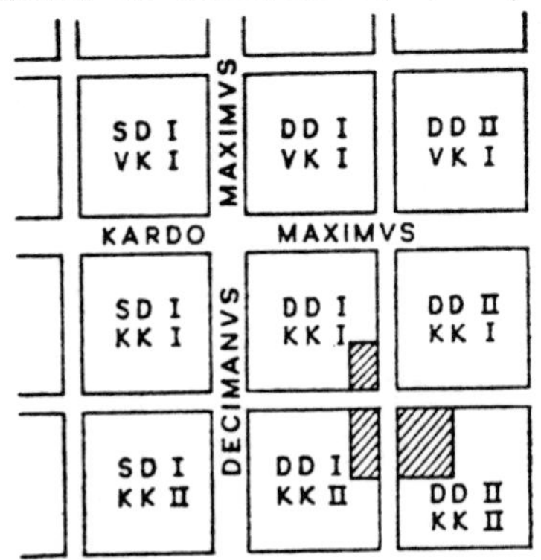

Fig. 14. Centuriation: the intersecting main roads (often abbreviated KM, DM) and the numbering of 'centuries' left and right of the *decumanus* (SD, DD) and on the near side of and beyond the *kardo* (CK or KK, VK). The shaded areas indicate three adjacent properties of a single owner, distributed between three 'centuries' (p. 88).

colonies, is a large part of Campania round Capua, whose land was declared *ager publicus* after it had sided with Hannibal. In north Africa the whole area of Carthaginian territory corresponding to much of northern Tunisia was centuriated after the defeat of Carthage in 146 BC; while an enormous area to the south of it was similarly divided, at least in skeleton, after the revolt of Tacfarinas had ended with his death in AD 24. An area hardly explored for centuriation until fairly recently is Spain; the Agrimensores refer to Mérida and other areas, and now quite a number of sites have been attested.[12] In the other western provinces, whereas centuriation is commonly found in southern France, evidence is sparse elsewhere. This may be because Mediterranean climatic areas were more favoured; moreover they do not experience the deep ploughing of countries such as England and the Low Countries. Nevertheless in these areas several undoubted and several likely centuriation schemes have now been established.[13] The sizes and shapes of 'centuries' are sometimes recorded in the Corpus Agrimensorum; in Livy many allocations of land to settlers are recorded, and these can often be tied up with 'century' size; while archaeology and aerial photography enable researchers to record remains of centuriation, sometimes of different periods superimposed at different orientations. The most ambitious record of the centuriation of a province is the *Atlas des centuriations romaines de Tunisie*.[14]

In addition to the text of the Corpus Agrimensorum, we are fortunate in possessing many miniatures, which are in the form either of paintings or of geometrical drawings. They are mostly inserted after the appropriate part of the text, though one set appears as a self-contained unit. The earliest preserved are in MS A (Arcerianus A, Wolfenbüttel), now thought to date from about AD 500;[15] while those in P (Palatinus latinus 1564, Vatican) are of the ninth century. In both, the colours are well preserved and the wording on the whole very legible. We can deduce from the text that some of the treatises in the Corpus were illustrated from the start. To what extent the miniatures represent original illustrations is discussed at the end of this chapter. The respective reliability of A and P may perhaps best be assessed from the large maps mentioned below (p. 97).

Clearly the maps and plans in the Corpus were designed not as official documents but as teaching maps;[16] this is indicated not only by their appearance but by some of the wording, e.g. the very general descriptions of colonies; the use of *ut*, 'for example'; and the introduction of famous names from the past, like Scipio. Nevertheless, the presence of so many indicates that either the *agrimensores* or the compilers of the Corpus or both were very map-minded. Since they were giving instruction in making plans, such large-scale maps could be more scientific and accurate than early world maps. (1) They needed to map a surveyed landscape, which for official purposes should essentially have been in plan. In the Corpus the pictorial aspect is introduced for demonstration purposes, somewhat as we might have used a ground photograph alongside a map in the days before aerial photography. The surveyed landscape is extended from local areas to colonies with surrounding territory. In some cases plan and elevation are awkwardly combined. (2) Demonstration maps were sometimes required for legal cases, and teaching maps were devised to illustrate how these should be compiled. Such illustrations could be either diagrammatic or pictorial or both. (3) Geometric drawings are sometimes given to show how to deal with irregular land, e.g. such *subseciva* (odd pieces of land) as lay between the centuriated land and the boundary of surveying. The use of triangles is very limited: if squares and rectangles would suffice, triangular measurements were not advocated; but we have several illustrations showing the relevance of the theorem of Pythagoras or of similar triangles. (4) The illustrations of a cosmographical nature are concerned either with planetary circles in a geocentric universe or with theoretical aspects of orientation.

We may classify the miniatures, for their contribution to cartography, into (a) those which are definitely maps or plans, (b) those which have a cartographic element, (c) others. Under section (a), the most numerous and striking are those which represent a colony with its territory. Since in each case the purpose is didactic, connected with the training of surveyors, rather than topographical, we have a variety of approaches from the topographically identifiable to the purely general designed as an example. The two which are most convincingly portrayed are Anxur-Tarracina (Terracina) and Minturnae (Minturno scavi).[17] Tarracina was founded in 329 BC on lands of the Volsci, each of the three hundred settlers receiving only 2 *iugera* (1.246 acres = 0.504 ha.), no doubt supplemented by common pasture. The Via Appia, originally laid out in 312 BC, led directly to the colony. In MS P of the Corpus Agrimensorum the illustration explains the text of Hyginus Gromaticus: 'In some colonies they set up the *decumanus maximus* in such a way that it contained the trunk road crossing the colony, as at Anxur in Campania. The *decumanus maximus* can be seen along the Via Appia; the cultivable land has been centuriated; the remainder consists of rugged rocks, bounded as unsurveyed land by natural landmarks.' The diagram shows, to left, the Pomptine marshes (*paludes*) and the Via Appia crossing them: the mountain range, the centuriation, the walled colony (*colonia Anxurnas*) and the sea. The only slightly incorrect detail concerns the position of the centuriated land. We know from the layout of farm tracks that there was centuriation in the area now called La Valle between the Via Appia and the mountain ridge; there probably was also, in the area where the diagram shows it, between the Via Appia and the sea, where repeated changes of land use may have resulted in the traces being obliterated.

Minturnae in Campania is likewise represented by a clearly recognizable miniature (pl. 10) of MS P. Originally an Ausonian settlement, it was colonized by Rome in 295 BC and restored by Augustus with territory on both sides of the R. Liris (Garigliano).[18] The accompanying text by Hyginus Gromaticus tells us: 'Augustus also re-founded a number of cities previously founded as colonies but depopulated in the civil wars, by sending out new settlers and sometimes increasing the territory. The result is that in many areas new centuriation cuts into the old at a different angle, the stones at the old points of intersection still being visible. For example, in the territory of Minturnae in Campania, the new assignation on the other side of the Liris is centuriated; on this side of the Liris a later assignation was made on the returns of the previous occupants (*possessores*).[19] Here the original boundary stones have been abandoned.' The miniature looks less correct than that of Tarracina; for example, the bronze statue (*aena*) is unlikely to have been outside the city. Moreover the walls enclosed a rectangle rather than a polygon as in the miniature.

A third example is even less specific: Hispellum (Spello: the 'Hi-' has already disappeared when the manuscript was written) in Umbria. Here the text of Hyginus Gromaticus reads: 'Many founders (of colonies) looked to suitability of terrain, and set up their *decumanus maximus* and *kardo maximus* where they were going to make the most of their assignation of land. Men of old, because of sudden dangers of war, were not content to wall cities, but also chose rugged, hilly land to provide a natural defence. Such rocky areas could not be centuriated, but were left either as state forests or, if barren, unoccupied. To bring the land of such cities up to the size required for colonies, they were given territory of neighbouring communities, and the *decumanus maximus* and *kardo maximus* were set up on the best soil, as in the territory of Hispellum in Umbria.'[20] The wording on the miniature, Colonia Iulia, is likely to be correct in that Hispellum was founded as a colony probably by Octavian. Schulten thought of the *flumen finitimum* as the R. Ose, which formed a boundary with the territory of Asisium (Assisi).[21]

In addition to these topographically identifiable colonies, there are miniatures of a number of others, either with no inscription or with a name which does not help with identification. Where, as most often, the KM and DM are outside the city, their meeting-point is normally shown in the Corpus as a T-junction rather than an intersection. This is perhaps connected with the original size of illustrations, though as they stand, most could have been extended downwards. Hyginus Gromaticus implies that the meeting of the KM and DM at the centre of a settlement was typical of military forts, quoting Ammaedara (Haïdra, Tunisia) as an example. So even though the diagrams labelled (a) *colonia Augusta* and (b) *colonia Claudia* in the land of the Tegurini may refer to Aosta and Avenches respectively,[22] it is not at all likely that the KM and DM of the centuriation intersected in the middle of these colonies. By the second century AD they were civilian settlements, and we know enough of their planning to judge that these diagrams are not true representations of them.[23]

The surveying of boundaries formed an important part of the duty of *agrimensores*, and several plans illustrate this. One diagram in MS P (fig. 15) shows squares of surveyed land, then an 'excluded and unassigned area' (such areas were often, though not here, called *subseciva*), and outside this the boundaries marked *fines Iuliensium*, *fines Mantuanorum*, i.e. 'territory of Colonia Iulia', 'territory of Mantua'. The wording of Hyginus Gromaticus is: 'An excluded area is so called because it is shut out beyond the centuriated area by a line forming a boundary. If the line forming the boundary is not surrounded by centuriation, it will be best to

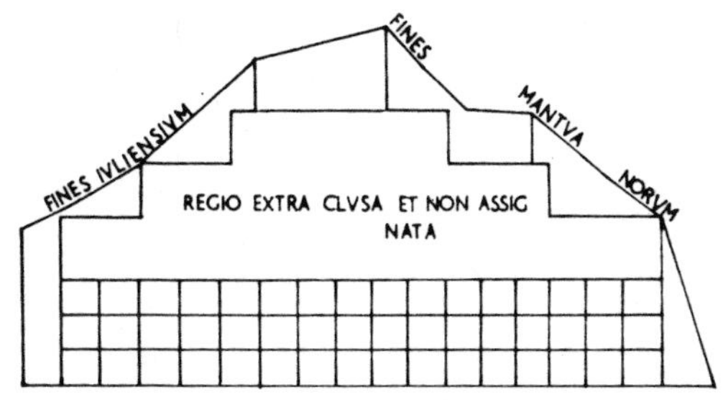

Fig. 15. Centuriated area and area excluded from centuriation and unallocated: miniature illustrating the text of Hyginus Gromaticus.

square off the end of the land with right angles surveyed with the groma, and put boundary stones in that way.' Colonia Iulia is a sample colony, while mention of Mantua sounds like a literary reminiscence. After the battle of Philippi, 42 BC, the land commissioners headed by C. Asinius Pollio allotted the lands of Mantua to discharged veterans. Virgil's father thus lost his lands at Andes near Mantua,[24] but seems on appeal to have obtained recompense.

Boundaries were an important element in land survey. One diagram (pl. 12), as illustrated in MS A, shows a triangular boundary stone, in the form of an altar, in semi-oblique view, at a point where the boundaries of three territories met. This *trifinium* is symbolically enlarged to emphasize its importance. Hyginus Gromaticus' text reads: 'But in certain areas we shall have to place stone altars, whose inscription shall show on one side, facing the centuriated area, the boundary of the colony, on the other, exterior side, the neighbours. Where the boundaries form a corner, we shall place triangular altars; similarly in mountainous areas.'[25] The Latin inserts the word *ut*, here 'for example', between *fines* and the genitive plural, thus showing that this is not an extract from an actual map but a demonstrative example. That can be virtually proved, since the two places out of three which are identifiable, Falerio (or possibly Falerii) and Vetona, are rather too far apart to have had a common boundary.

Within a centuriated area the *limites* which separated the squares or rectangles had to be kept clear. Thus Frontinus writes: 'Road disputes are covered in lands called *arcifinii* by ordinary law, while in allocated land they are covered by means of measurements. All dividing strips (*limites*) should, according to the law dealing with colonies, serve a public right of way; but many of them naturally go through steep, rough terrain, where no road can be made, and are used as part of the local fields, where the landowner, whose wood may happen to embrace the dividing strip, may inconsiderately refuse a right of way.'[26] The diagram illustrating this may be a later addition; it tries unsuccessfully to combine plan and elevation. At the top is a mountain in elevation; below is centuriated land in plan, with trees in elevation and farmhouses (*villae*) semi-oblique. As a result, the *limes* climbing the mountain is shown at an oddly acute angle. Somewhat similar is the miniature of Mt Mutela in the Sabine country (location unknown) with a farm at its foot.

Other sketches of farmlands may also include farmhouses. In a Frontinus illustration the object of the miniaturist seems to have been to include as many of the types of boundary mentioned in the text as possible. Frontinus comments: 'Land known as *arcifinius*, which is unmeasured, is bounded, following long-standing observance, by rivers, ditches, mountains, roads, trees in front, watersheds or ground of which an earlier holder could claim possession. Its name, according to Varro, is derived from keeping off (*arceo*) the enemy.'[27] In the illustration the small spaces on the river bank represent boundary stones in ground-plan.

Of particular interest to the cartographer are such illustrations as deal with the compilation of survey maps. Normally a map (*forma*) was drawn up in two copies, one kept in the *tabularium* in Rome, the other in the local community. The land surveyor had to be taught to enter on them the areas, the legal status and the names of occupants, with number of *iugera* held; also such features as mountains, rivers, marshes and roads. The details of all these come from Hyginus Gromaticus. (1) 'When we have ended all "centuries" with inscribed stones, we shall surround parts assigned to the State, even if they are centuriated, with a private boundary, and shall enter them on the map appropriately, as "public woods" or "common pasture" or both. We shall fill the whole extent with the inscription, so that on the map of the area a more scattered arrangement of lettering may show greater width. . .'.[28] The wooded area surrounded by centuriated land is labelled *silva et pascua publica*. Whereas the chief concern was legal exactitude, the comment on lettering seems to show that some training in map-making was given.

1 Fragmentary plan of Nippur, Babylonia, on a clay tablet now at Jena. For recent theory on its orientation and coverage see McG. Gibson in *Archaeology* (USA), 30 (1977), 34–7.

2 The Babylonian World Map, *c.* 600 BC.

3 *Opposite*, fresco from the tomb of Menna at Sheikh Abd el Qurna, Thebes, Upper Egypt. In the upper strip, surveyors with ropes and other measuring equipment; in the lower, an official and scribes. Borchardt attributes it to the reign of Amenophis IV.

4 Fresco from the 'house of the admiral', Akrotiri, Santorin, Greece, *c.* 1600 BC, showing a seaside settlement with map-like features.

5 The Novilara Stele, a tablet from Novilara (Pesaro-Urbino), probably of the 7th century BC, with ships and features which may be in plan.

6 Obverse of Roman portable sundial, 3rd century AD, on which the pointer can be set to latitude (XXX–LX) and to date (VIII K IAN – VIII K IVL). The gnomon throws a shadow on a curved band marked with Roman 'unequal' hours.

7 The celestial globe of the Farnese Atlas, 1st–2nd century AD copy of a Hellenistic original.

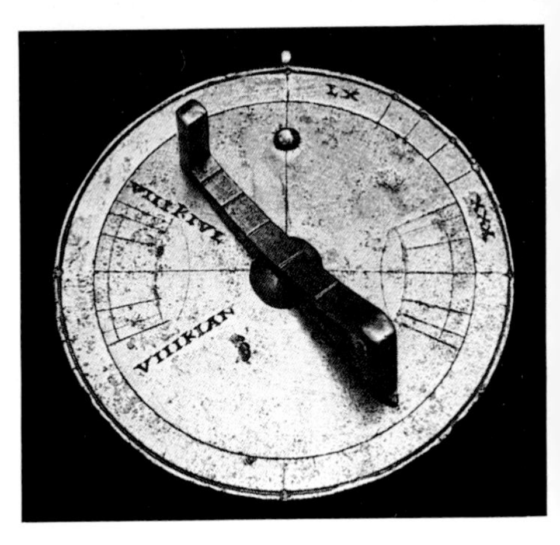

9, 10 *Opposite*, miniatures illustrating the text of Hyginus Gromaticus. *Top*, plan of Anxur-Tarracina (Terracina): colony (COLONIA AXVRNAS = Anxurnas), sea, Via Appia, centuriation, Pomptine marshes (PALVDES), mountains. *Bottom*, plan of Minturnae (Minturno Scavi): colony (MYNTVRNAE), sea, R. Liris (Garigliano), mountains, bronze statue (AENA), new allocation of lands (ASSIGNATIO NOVA).

8 Iron end-pieces of a wooden measuring rod, *decempeda* (10 Roman feet long) or *pertica*, found at Enns, north Austria.

bus continetur. terminata in extremitate more arci
finio per demonstrationes et locorum uocabula.

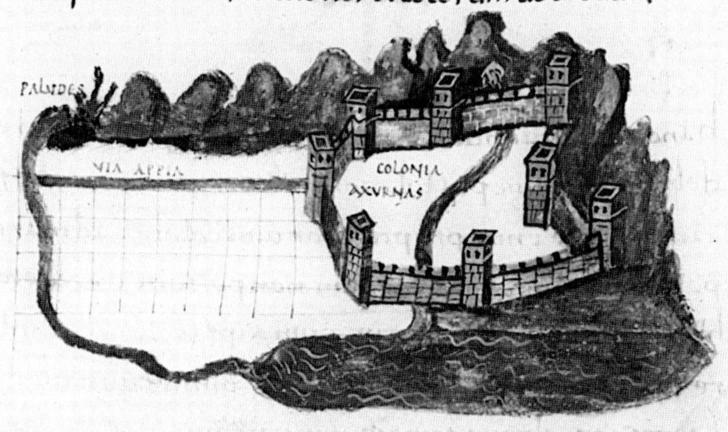

Quibusdam coloniis postea constitutis sicut in africa
admedere .D. m. et K. ex his a ciuitate oritur et per quit

finium commutatione relictis primae assignationis
terminis more arcifinio possidetur,

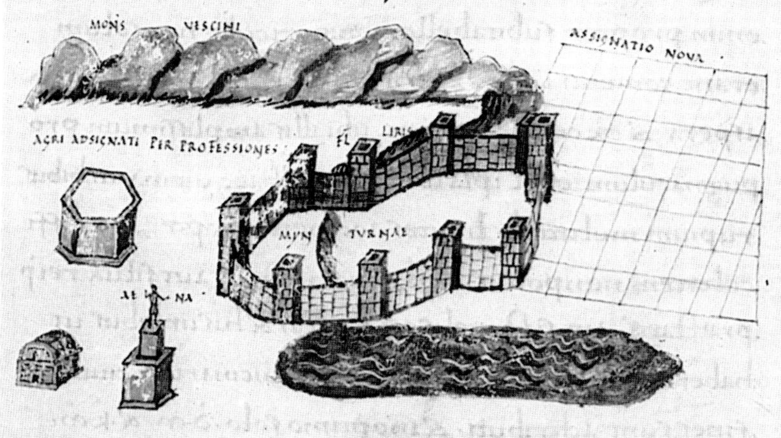

Multis ergo generibus limitum constitutiones in
choatae sunt. quibusdam coloniis. K. m. et D. m. non

limites primos nisi a foris accipere non possunt.

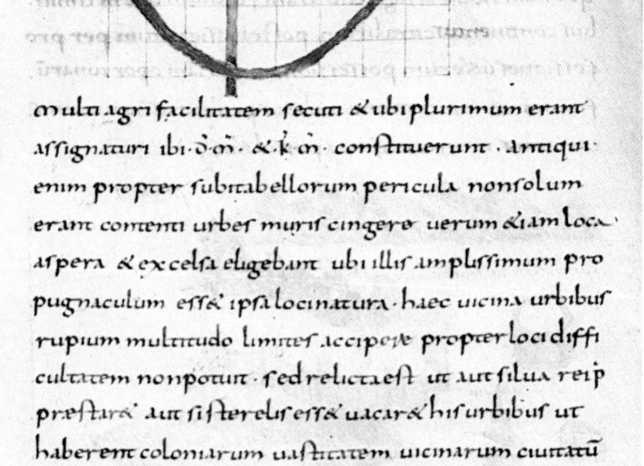

Multi agri facilitatem secuti & ubi plurimum erant
assignaturi ibi d. m. & k. m. constituerunt. Antiqui
enim propter subita bellorum pericula non solum
erant contenti urbes muris cingere uerum etiam loca
aspera & excelsa eligebant ubi illis amplissimum pro
pugnaculum esset ipsa loci natura. haec uicina urbibus
rupium multitudo limites accipere propter loci diffi
cultatem non potuit. sed relicta est ut aut silua reip
praestaret aut si sterelis esset uacaret his urbibus ut
haberent colonarum uastitatem uicinarum ciuitatu(m)
fines sunt adtributi. & in optimo solo d. m. & k. m.
constituti sicut in umbria finibus ispellatium.

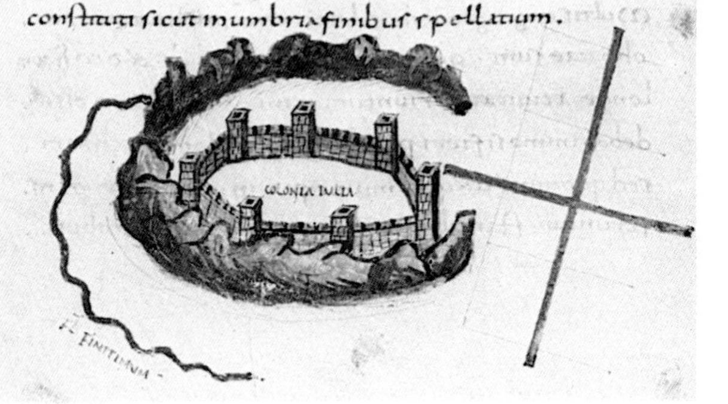

RAS ERIT AD FINES UBI FINIS ET ANGULUM FACIET TERNAS
60 ANGULU RUG ARAS PONE MODUS SIC ET IN LOCIS montuosis

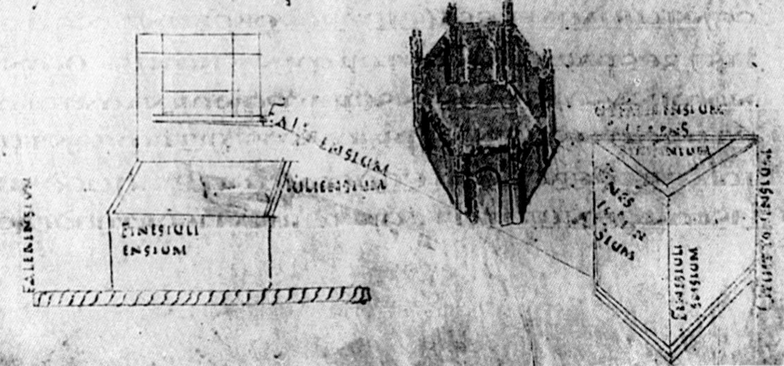

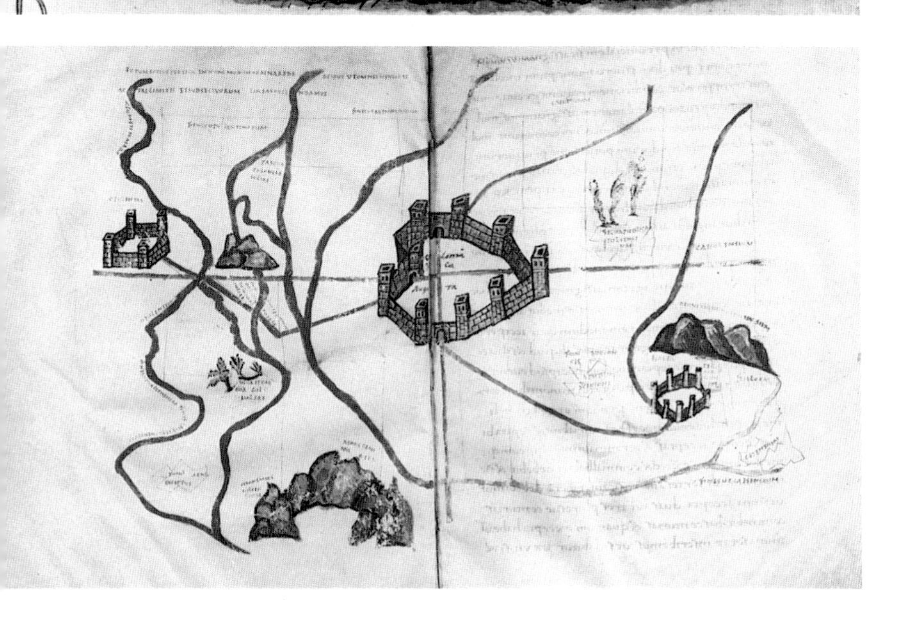

1 *Opposite (top)*, further miniatures
lustrating the text of Hyginus Gromaticus
nd showing colonies with centuriation
riginating outside the settlement. In the
ower one may be seen the Colonia Iulia
= Hispellum), a river acting as boundary
f the territory (FL. FINITIMVM), and *cardo*
nd *decumanus* of centuriated land.

2 *Opposite (bottom)*, miniature (Frontinus)
howing *trifinium*, boundary between three
erritories. The word *ut* indicates that the
ames of inhabitants of colonies or
unicipia are intended only as examples.

13 *Top*, farm boundaries (*F* = *fundus*).
Miniature illustrating the text of Frontinus,
De controversiis, on boundary disputes.

14 *Bottom*, map of several settlements, with
centuriation, mountains etc. Miniature
illustrating the commentary of Agennius
Urbicus: Colonia Iulia Augusta (Augusta
Taurinorum, Turin); to right, Hasta (Asti);
to left, Opulentia, not identifiable. A
reference to the territory of Segustero is
thought to be a mistake for Segusio (Susa,
north Italy).

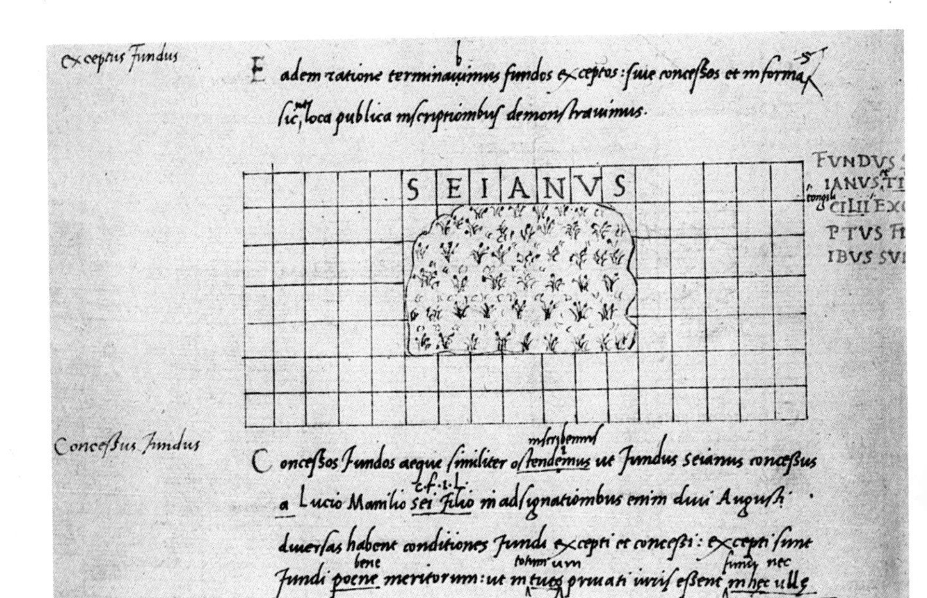

15 Miniature from the Corpus Agrimensorum: an 'excepted' farm, one not included in surrounding centuriation.

16 Plans of farm types designed round Greek letters.

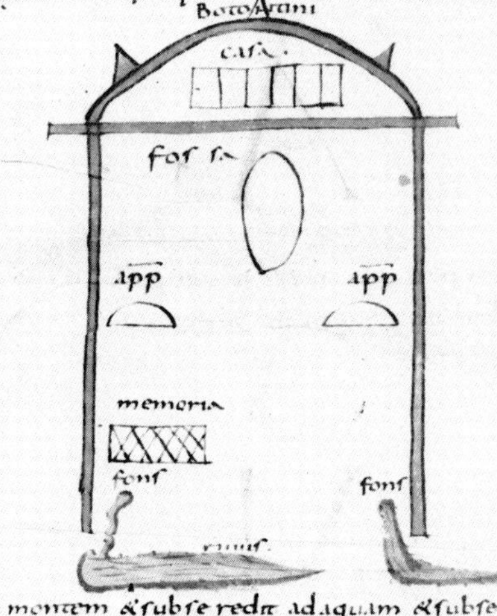

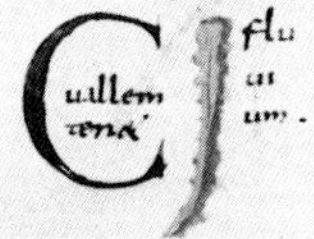

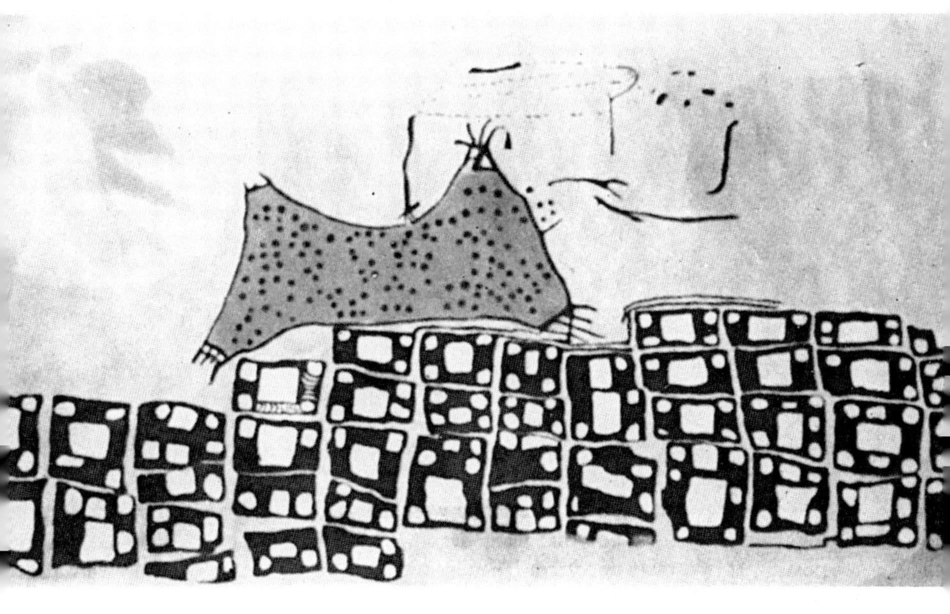

17 Town plan from Çatal Hüyük, central Anatolia, with volcano in perspective, *c.* 6200 BC.

Per longum currit ei limis ante se hab& casale post
se ad pedem hab& aquam uiuam & flumen inferius.

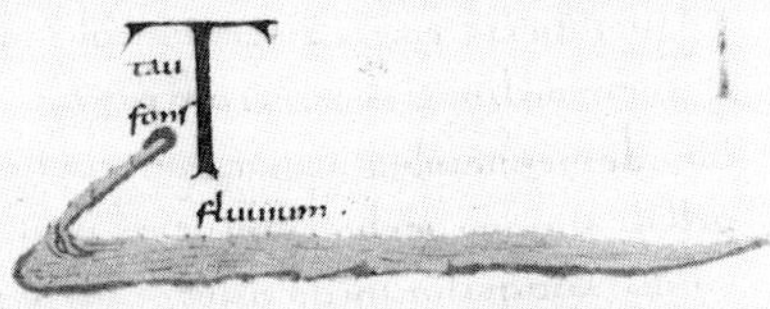

n montem se colligit in octogonum iac& per mediū
flumen hab& & ad pedem aquam uiuam & flumen in
erius.

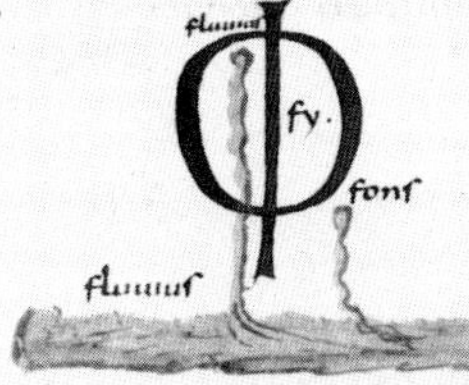

plano contra pectus iac& & sub se mord& eam ·
lia casa dextra leuaq; aquas uiuas hab& & flumen in
erius.

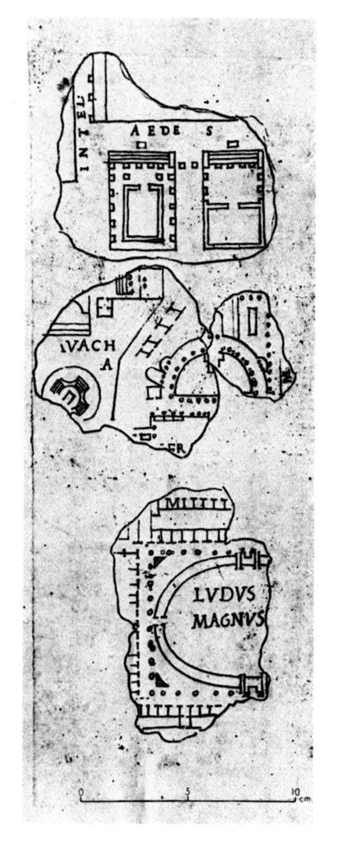

18, 19 Forma Urbis Romae. *Above*, an area of the Campus Martius, including the shrine of the deified Vespasian and Titus, the Saepta Iulia and the *diribitorium*, a small building where votes were counted. *Left*, drawings on a Vatican manuscript of fragments found during the Renaissance. The Ludus Magnus was a gladiatorial school near the Colosseum. The adjacent fragment serves to supplement those found more recently and shown in pl. 18 above.

21 The Pesaro anemoscope: the so-called Boscovich anemoscope, *c.* AD 200, found near the Porta Capena, Rome, and now at the Museo Oliveriano, Pesaro. It is a thin cylinder of marble with central hole for pole supporting pennant and small holes near the rim for wooden pegs indicating the winds.

< 20 Inscription of Arausio (Orange), AD 77, setting up a cadaster by order of Vespasian. The restored and expanded Latin text reads: Imperator Caesar Vespasianus Augustus, pontifex maximus, tribunicia potestate VIII, imperator XVIII, pater patriae, consul VIII, censor, ad restituenda publica quae divus Augustus militibus legionis II Gallicae dederat, possessa a privatis per aliquod (= aliquot) annos, formam proponi iussit, adnotato in singulas centurias annuo vectigali, agente curam . . Ummidio Basso proconsule provinciae Narbonensis.

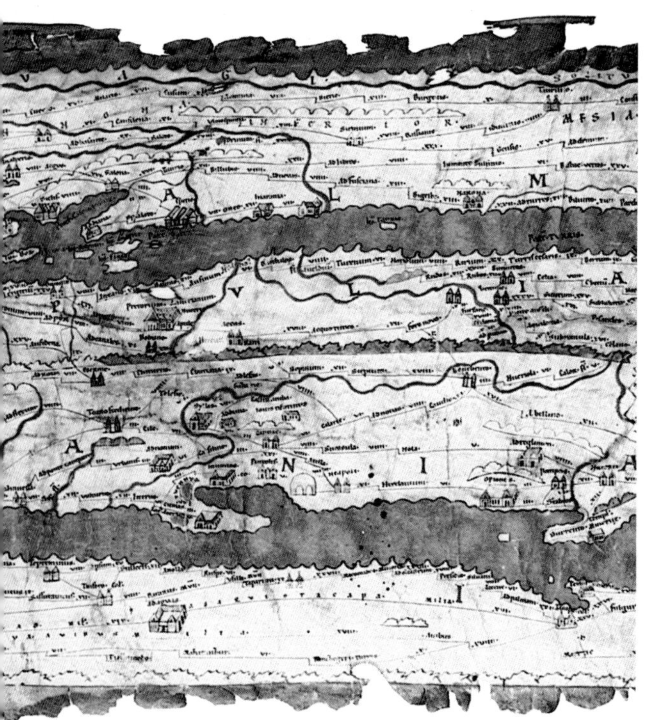

24 Ostia, mosaic from the 'Forum of the Corporations', AD 197–200, probably depicting the Nile delta and a pontoon between Memphis and Babylon (Old Cairo).

:, 23 *Opposite*, the Peutinger Table.
op, the section containing part of
:ythia, Macedonia, Thrace, Epirus,
ainland Greece and an area now in
ibya. The Peloponnese, though
ongated like the remainder, is more
refully delineated than many areas.
ottom, the section containing central
urope, Dalmatia, Apulia, Campania and
ırt of north Africa.

5 The Madeba Map: the area round
richo (Ιεριχω, left), shown with a grove
f palm-trees, and Jerusalem (part
ıown bottom right).

› World map accompanying Book VII
f Ptolemy's *Geography*. As in all
tolemy manuscript maps except the
/ilczek-Brown Codex (L 40), east Africa
joined to south-east Asia by *terra
cognita*.

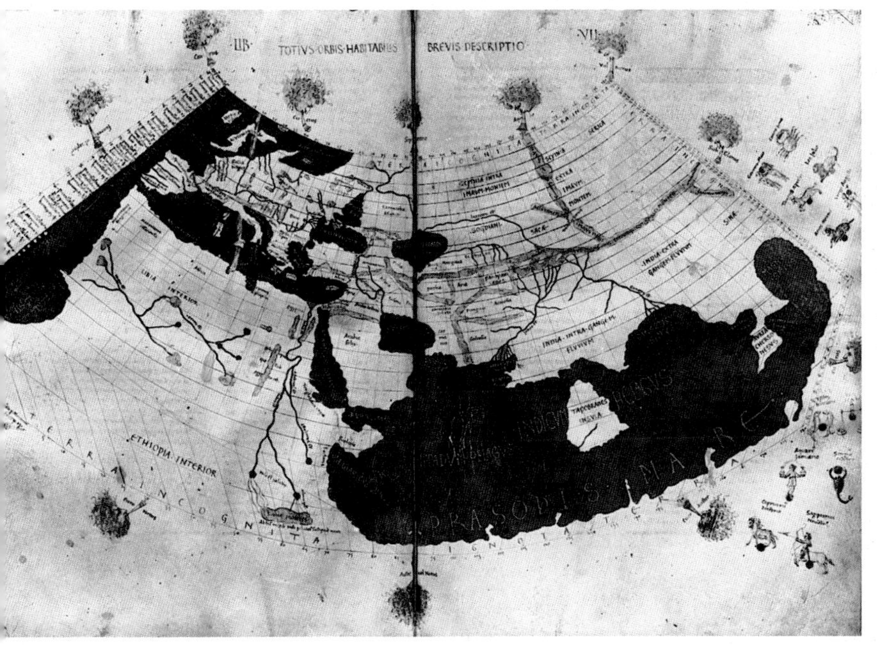

27 Southern Italy, regional Ptolemaic map, Vatican, late 13th century. A notable feature, as in Ptolemy's co-ordinates, is the bend in the 'heel' of Italy.

30 *Opposite*, plan (or projected plan) of St Gallen monastery, Switzerland, AD 818–37, with east at top. Centre is the cloister, to north the long chapel with double apse, to south the refectory. The scriptorium is north of the east apse. There are also an infirmary, a noviciate, a school, an abbot's house, a guest-house, a cemetery, kitchens and other service areas, and animals' quarters.

28 The British Isles, regional Ptolemaic map, early 15th century. The most notable feature, as in the co-ordinates, is the orientation of north Britain. Each *polis* is allocated by a symbol (cf. fig. 28) to its appropriate tribe. Urolanium (Verulamium) and Vinnovium (Binchester) are incorrectly located. This manuscript is closely copied from Urbinas graecus 82.

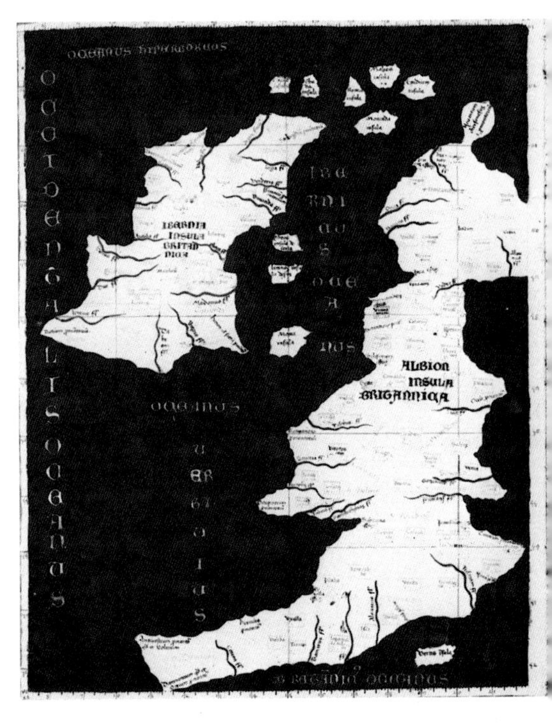

29 Miniature world map illustrating the *Prokheiroi kanones*, 'handy tables', an astronomical work of Ptolemy's; the map itself may have originated under the Late Empire. In the rectangle, reading downwards, are places in Egypt: lower country, seven-nome area, Syene (Aswan: at intersection of axis and Tropic of Cancer), Hiera Sykaminos, Lake of Meroe. In the upper semicircle, Persian Gulf of the Erythraean Sea; in the lower sector of the circle, Hades and its rivers.

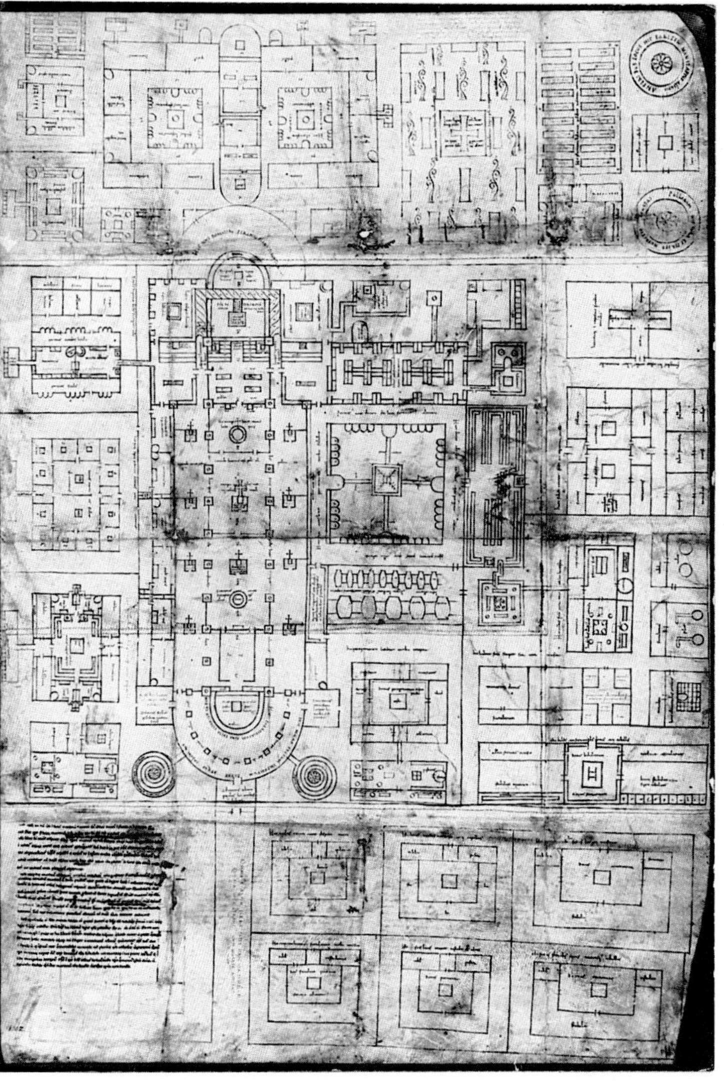

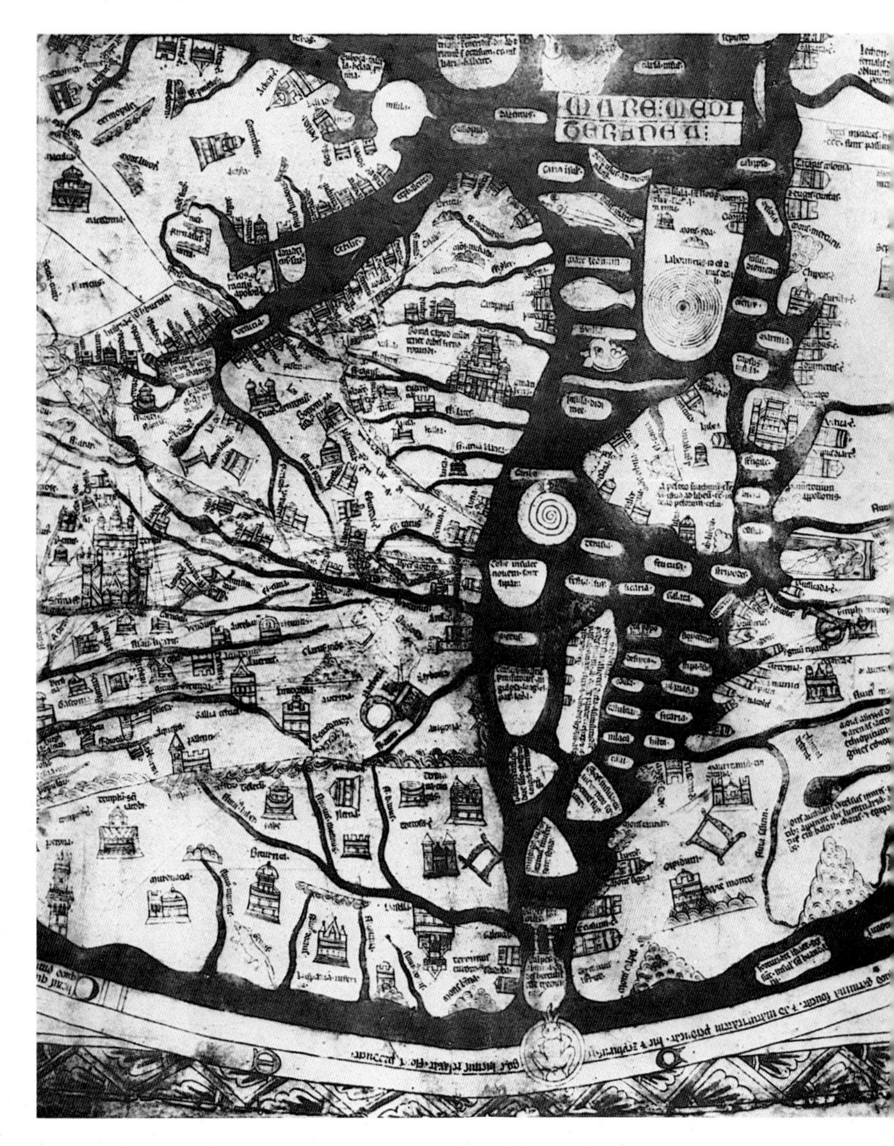

31 The Hereford World Map, *c.* 1300. The Mediterranean (MARE MEDITERANEŪ), with a large equilateral Sicily occupying its centre, and also featuring marine life.

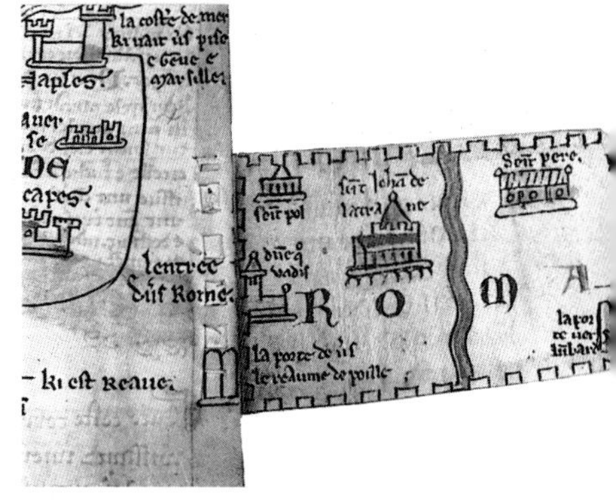

32 The Matthew Paris pilgrimage map: the portion showing Rome and the road to Naples. The Rome city plan, with the Tiber and St John Lateran and other buildings, is dovetailed on.

(2) 'We shall similarly show granted (*concessi*) farms, e.g. "Farm of Seius granted to Lucius Manilius son of Seius". In Augustus' allocations of land, excepted farms have a different status from granted farms. Excepted farms were awarded for good service and are subject only to private law. . . . Granted farms are the property of persons allowed to possess more land than the edict permitted . . . . We shall inscribe them "granted" so that they may remain thus on the map.'[29] The translation of the wording on the miniature is: 'Farms of Faustina as originally granted by P. Scipio'. The allusion to the ownership may be to Faustina (d. AD 140–41), empress of Antoninus Pius, or to his daughter of the same name. But the rest is either purely general, Seius serving as a sample name in legal writings, or archaic, referring to one of the famous generals of some centuries earlier.

(3) 'We shall write both on the maps and on the bronze tablets all mapping indications, "given and assigned", "granted", "excepted", "restored, exchanged for own property", "restored to previous owner", and any other abbreviations in common use, to remain on the map. We shall take to the Emperor's record office the mapping registers (*libri aeris*) and the plan of the whole surveyed area drawn in lines according to its particular boundary system, adding the names of the immediate neighbours. If any property, either in the immediate neighbourhood or elsewhere, has been given to the colony, we shall enter it in the register of assets. Anything else of surveying interest will have to be held not only by the colony but by the Emperor's record office, signed by the founder.'[30] The *libri aeris* were registers, at first no doubt on papyrus rolls, then on parchment codices. The Emperor's record office, *tabularium*, was on the slope of the Capitoline Hill in Rome, where its substructures are well preserved. The illustration appears in very different forms in A and P. Whereas A has no wording and three intersecting straight lines at right angles, indicating three centuriation schemes on the same orientation, P shows in the centre Colonia Iulia Augusta (Turin), and to the right not an adjacent town but Hasta (Asti), actually 50 km ESE of Turin; a place Opulentia, non-existent as such and misleading if corrupt for Pollentia (Pollenza); and an unhelpful reference to Segustero.[31]

(4) 'This is how we shall allocate undeveloped land in the provinces. But if a borough has its status changed to that of colony, we shall examine local conditions and allocate the land as these

conditions require. . . . This land we shall allocate, according to the specified law or if we wish according to Augustus' law, "as far as the scythe and plough have gone". This legal phrase is thought by some to refer only to cultivation; but I think it means that all useful land should be allocated. . . . You will inscribe the lots in such a way that if a single holding extends over two or three or more "centuries", these "centuries" shall be inscribed in one single lot with the amount of land held in each. For example, if 66⅔ *iugera* are given to one man and are split between three "centuries" thus, DD I KK I 6⅔ *iugera*, DD I KK II 15 *iugera* and DD II KK II 45 *iugera*, a single lot will have to embrace these three. The rest will be carried out according to this example. We shall take the owners of lots to their land and assign boundaries to them. . .'.[32] The illustrations in A and in P do not tally well with the text. In the sixteenth-century MS Vaticanus latinus 3132 we have two centuriation schemes side by side, the latter representing a *municipium* which has been given colony status. None of the manuscript miniatures shows the detail of a holding split between three 'centuries', which can however be reconstructed. But it looks as if A and P, in different ways, misinterpreted the phrase as if it referred to three systems of centuriation. Then the ninth-century scribe of P, feeling the need of a regional map, inserted winding main roads (*viae consulares*) and places from different areas of Italy, resulting in an incoherent attempt at a map.

Further instructions to surveyors are included in the manuals: (1) Geometrical problems in the Corpus Agrimensorum are illustrated by diagrams, but few of these are map-like. However, M. Iunius Nipsus, *Fluminis varatio*, has a text which reads: 'If as you are squaring land you have a river in the way which needs measuring, do thus. From the line perpendicular to the river, make a right angle, placing a crossroad sign. Move the groma along the line at right angles, and make a turn to the right at right angles. Bisect the line from this right angle to the original one, and place a pole at the mid point. With the groma here, make a line in the opposite direction from the pole you had placed on the other side of the river. Where this meets the line drawn at right angles, place a pole, and measure the distance from here to the crossroad sign. Since you have two triangles with equal perpendiculars, their bases too will be equal. So you now know the distance from the pole across the river to the first crossroads sign; subtract the distance

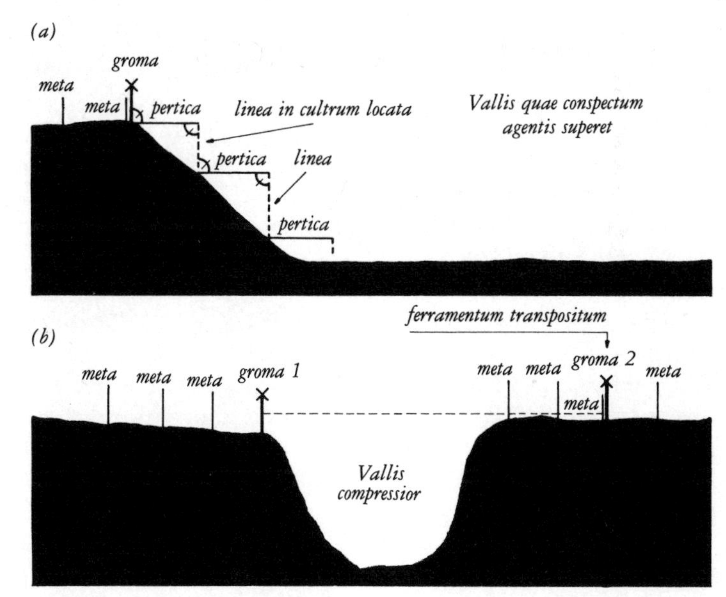

Fig. 16. Two sections reconstructed by N. Alfieri to show the way in which the Roman land surveyors may have practised *cultellatio* (levelling): (a) in a wide valley, (b) in a narrow one.

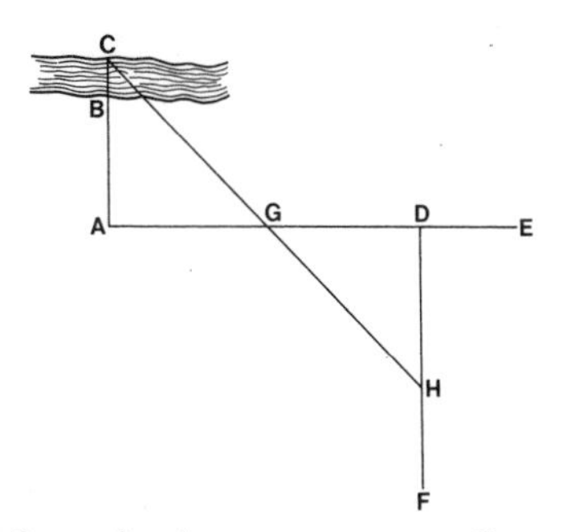

Fig. 17. Geometrical figure for river measurement: Corpus Agrimensorum, Nipsus, *Varatio fluminum* (redrawn). To find width of river, BC: since the triangles CAG, HDG are congruent, DH = CA; subtract length of AB from length of DH.

from the latter to the river, and you have the width of the river.'[33] The accompanying diagram is mathematically inaccurate; a modern reconstruction has been attempted. (2) Orientation exercises in survey operations show that both the sundial and observation of the sun's position were used. An unsatisfactory combination of the ground-plan of a gnomon with that of centuriation and a sketch of a mountain in elevation with the sun in two positions, on each side of it, are among the manuscript illustrations. Hyginus Gromaticus' alternative method of obtaining south has been shown to depend on solid geometry;[34] but this with its accompanying diagram is quite exceptional. Observation of the stars may also have been practised for this purpose, but is not specifically mentioned in the text. (3) There are also simple didactic illustrations involving celestial cartography, to illustrate the knowledge of cosmology that land surveyors were supposed to acquire.

Fixed orientation in drawing the plans is unlikely. The entry to a property, as with modern plans, or an important feature like the sea in the case of a colony with centuriation, is likely to be at the bottom. However, since there was a distinct tendency to orientate surveys by the cardinal points, unless relief (or a pronounced feature like the Via Aemilia in the Po valley) dictated otherwise, north tends to be at or near the top. It will be seen, however, that the Forma Urbis Romae has roughly south-east at the top, and such orientations are not as rare in classical times as is commonly supposed (p. 104).

The miniatures of the Corpus Agrimensorum are thus a somewhat heterogeneous collection, as regards dating, style and content. A treatise may have been written in the late first or second century AD, its miniatures may have been added under the late Empire, and the whole re-edited to form a Corpus about AD 500. It is inevitable that discrepancies between manuscript illustrations and the text, or between one manuscript and another from that time onwards, should occur. The style of execution has much in common with that of fifth/sixth-century Byzantine illustration, though the manuscripts we have are likely to have been produced in Italy. The cartographic content is very uneven, varying between areas where there is realistic detail, indicating some local knowledge, and purely general model illustrations which could only be used for general teaching purposes.

Finally the question remains whether the two main authors quoted in the above extracts, Frontinus and Hyginus Gromaticus, are likely to have used illustrations in their land-surveying treatises. As regards the second, we cannot say, since he gives no indication in his text, which seems to be of the second century AD. Of the man we know only that he was not Augustus' librarian C. Iulius Hyginus, and he does not seem to have been the same as the Trajanic Hyginus who wrote *De condicionibus agrorum*. But in the case of Frontinus we can be more positive. Sextus Iulius Frontinus, born about AD 30, became governor of Britain for three years from AD 74 or 75. His work on surveying, together with his treatises on military science, was probably written about AD 85–90. In AD 97 he was appointed by Nerva curator of the waters of Rome, and under Trajan he wrote the work on Rome's water supply which has come down to us. In this he tells us that he set up maps of the aqueducts which included such details as height of piers, thus enabling any repair work to be carried out far more efficiently.[35] Even though this treatise was almost certainly written later, it is quite likely that even when Frontinus was writing on the technicalities of land surveying he considered it necessary for clarity to illustrate features in his text. Unfortunately it has to be admitted that for a Roman senator to take such a keen interest in technical details was only too rare. The Graeco-Roman educational system of the period caused greater esteem to be accorded to literary composition.

CHAPTER VII

ROMAN STONE PLANS

PREDECESSORS

The idea of a town plan obviously developed very early. The first extant example is not one carved on stone but a wall-painting from Çatal Hüyük, Turkey, which is now in the Museum at Konya.[1] Datable to *c.* 6200 BC, it represents the houses of an early settlement in oblique pictorial form, much like the *veduta perspettiva* of Italian Renaissance town plans. Although much earlier in date, it is more developed in technique than the rock carvings of Val Camonica in the province of Brescia, north Italy.[2] There are two periods of these, the middle Bronze Age and the Iron Age. The former bear some resemblance to a plan of a settlement, though they tend to confuse plan and elevation. Interpretation is not always easy, since there is no key; thus we do not know whether certain lines indicate paths or water-courses.

Since Greek town-planning of an orthogonal nature was so developed (p. 87), we may surmise that the planners worked either with models or with plans, but none of either category has survived.[3] It was a model, not a plan, of a proposed new city that the architect Dinocrates showed to Alexander the Great.[4] We do know the Greek technical term for a plan, *ichnographia*, literally the drawing of traces or tracks: Vitruvius mentions it as one of the skills which the perfect architect must possess.[5]

The only incised drawing likely to be a Greek plan so far discovered, at Thorikos, east Attica, has been mentioned on p. 26.

A POSSIBLE MAP OF GAUL

In 1976 a block of local sandstone, with maximum length and width 56 × 47 cm and with average breadth 14 cm, was found near the centre of the Roman camp at Mauchamp, near the R. Aisne.[6] It was taken by the finder to his house at Brie Comte Robert, where

Fig. 18. Val Camonica rock carvings: redrawing of map-like incisions at Bedolina, near Capo di Ponte, Val Camonica (Brescia). Those resembling ground-plans are of the Bronze Age, those with huts in elevation are of the Iron Age.

the present writer and his wife were able to examine it. There seems no reason why it should not, as the finder claimed, be a Roman map of Gaul, possibly associated with Caesar's campaigns. The side of the stone shows signs of chiselling and bears a resemblance to the west coast of France. It is also noticeable that, assuming this is a map, north appears at the top when the stone is stood upright. This is a common orientation of Roman maps, though by no means the only one (p. 100). The stone could have been worked by a military surveyor, perhaps for his own interest rather than for any official purpose.

There has so far been no conclusive proof of the date at which this stone was worked. If military surveyors were map-minded in this way, it is surprising that no other doodles of this sort have emerged. However, many interesting ancient stones have been overlooked or lost, as is particularly noticeable with centuriation and boundary markers.

TOWN PLANS

Of actual town plans we have in fact, apart from isolated fragments, only a partial reconstruction of the very large marble plan of Rome, dating from the early third century AD, which is mentioned below. But it seems likely that this had a predecessor on the same site, drawn up in the Flavian period: we are told that in 74 Vespasian and Titus had Rome measured.[7] Those responsible for drawing up such plans were presumably the *mensores aedificiorum*, literally 'measurers of buildings', whose activities also covered other aspects of urban survey.[8] Such a one was T. Statilius Aper, whose tombstone, showing in relief the implements of his trade, is in the Capitoline Museums, Rome.[9] We possess a number of their instruments, but unfortunately no manuals such as for land survey are found in the Corpus Agrimensorum. As a likely organizer of the scheme it is possible to think of Frontinus, who with his need to map the aqueducts of Rome (p. 101) may well have been interested enough to map the whole city: certainly aqueducts are shown on the plan.

The Forma Urbis Romae itself was completed some time between AD 203 and 208, in the principate of Septimius Severus.[10] We can be sure of this because it includes the Septizodium, an ornamental gateway erected in 203, and a building commissioned by the emperors Severus and Caracalla, which however from its lettering NN (*nomina*) was presumably awaiting completion.[11] The Forma Urbis was officially sponsored, but whether it was a new plan or a revision we do not know. It was in marble and originally measured 18 m 30 high × 13 m 03 wide. The fragments of the original are in the Palazzo Braschi in Rome, where a Spanish researcher has over many years been working on their reconstruction.[12] In a courtyard of the Capitoline Museums is a reproduction with copies of such extant pieces as were able in 1949 to be located with some certainty. Originally it was fixed to an outer wall of a library attached to Vespasian's Temple of Peace. That temple suffered from a fire in 191, and Septimius Severus restored it. Today, outside the Church of Saints Cosmas and Damian, the dowel-holes which helped to affix the plan to the wall can still be seen. Study of the positions of these holes has enabled archaeologists to work out where certain individual blocks went. Similarly some pieces were joined together in antiquity, and examination of the iron clamps or dowel-holes has been rewarding in the same way. Naturally the first pieces to be located were the inscribed fragments, but many of the others are now placeable. One of the earliest to be recognized (1590) was the Ludus Magnus, so inscribed, a gladiatorial school near the Colosseum set up by Domitian. Drawings of a number of fragments found in the Renaissance, but lost since, are preserved in a Vatican manuscript.[13] These have helped in attempts at reconstruction (pl. 19).

The top of the map can be identified because of the inscriptions: it is on average 43° east of south, with variants between 36° and 50°. It looks as if town plans were not governed by the conventions applicable to smaller-scale maps. Instead, this map seems to have been orientated to face roughly the way the public was looking, as do the Pesaro anemoscope and possibly the Orange cadasters.

Professor F. Castagnoli mentions that (a) when Augustus divided Rome into fourteen regions, the first region was in the south;[14] (b) according to Tacitus the gardens of Sallust were to the left part of the city;[15] and points out that south was a commoner orientation in antiquity than is generally supposed. The purpose of Augustus' division, with a subdivision of each region into *vici* (wards), was to improve municipal efficiency. Since Rome, which he boasted of having found built in brick and left built in marble,[16] was dear to his heart, we may surmise that he too commissioned some sort of map of the city, even if (to explain lack of mention) it was not publicly exhibited. As to the gardens of Sallust, on the Pincio, it would be possible by modern analogy to think of a left and a right bank, depending on position for one looking downstream. But there is no mention in antiquity of any such phrase, and if Tacitus was familiar with a Flavian map he could well have used it automatically.

The average scale is about 1:300, but the extremes in extant fragments are widely spaced between 1:189 and 1:413. It seems clear that the *mensores* started with public buildings of some size, and their 'technicians' were not as careful with the scales of these as they might have been. In any case any enlargement may have been intentional, so as to show public buildings as large and important. Scale is sometimes difficult to work out because a few features, chiefly aqueduct arches, are shown in elevation, not in plan. A possible explanation of this is that in plan they might have been difficult to recognize. Alternatively they might be considered as symbols, which like trees on Ordnance Survey maps would not be drawn to scale. Unlike public buildings, private houses have no wording or lettering on them. Staircases are represented either in plan or with one of two conventional signs: (a) a triangle with steps in plan inside; (b) two lines forming an acute angle and evidently standing for such a triangle.[17]

Apart from those already mentioned, extant names fully preserved include *porticus Liviae*, the colonnade completed on the Capitoline in honour of Livia in 7 BC; *inter duos pontes* for the Tiber island linked to each side by a bridge;[18] *aqueductium*, etc. Many other names preserved only in part have been able to be restored either certainly or probably, e.g. *(vicus) patricius*,[19] *(amphithe)atrum*, *(forum pa)cis*, *(basilica I)ulia*, *(circus m)ax(imus)*.

There are unfortunately many areas of Rome where the information available is very fragmentary. But one area where recent advances in research have been made is that of the Campus Martius between the Via delle Botteghe Oscure and the Pantheon.[20] Whereas under the Republic it was reserved for voting, exercise and army training, from Augustus' principate onwards it came to be thickly built up. Apart from the temple area of Largo Argentina, evidence from archaeology is only scattered owing to the density of modern settlement. Also the inscriptions on this part of the plan are incomplete and contain uncertainties. The largest building was a colonnade[21] of the deified Vespasian and Titus, measuring *c.* 180 × 75 m; the word DIVORUM ('of the deified men') is well preserved, and one can clearly make out the individual chapels to the two emperors. The building at first known only from fragmentary lettering was not a temple, as Lundström thought,[22] but Saepta Iulia, voting enclosures set up by Agrippa in 26 BC on Julius Caesar's plan and later adapted for a bazaar and gladiatorial shows. There are also a temple of the Egyptian god Serapis; a Porticus Meleagri, first recognized by Lundström; and a small triangular colonnade called Delta, first identified by Gatti[23] (earlier scholars thought it might be the Villa Publica). In the publication by Carettoni and others, another voting connection is included, the Diribitorium (only DIR preserved), a building where the votes were counted.

An assessment might therefore be that, although the scale is in places imperfect, and although from time to time archaeologists uncover sites of buildings which do not tally with the outline plans of the Forma Urbis,[24] this can nevertheless be claimed as the most accurate plan of Rome until that of G. B. Nolli in 1748.

Other remains of Roman town plans are only fragmentary.[25] For example, one in Rome includes names of private property owners, who are not mentioned on the Forma Urbis. Again, a fragment from Isola Sacra, near Ostia, although it has no lettering, has numerals which may well be measurements in Roman feet.

The tradition of town plans spread from Rome to Constantinople and may have continued after the fall of the Western Empire. Among the treasures bequeathed by Charlemagne[26] were gold and silver *mensae*, presumably table-tops, the silver ones engraved with representations (see below) of Rome, of Constantinople and of the world. Whether the first two were maps or pictorial representations is not quite clear.

Shape	*Type of representation*	*Subject*	*Legatees*
square	descriptio	Constantinople	Vatican
circular	effigies	Rome	Ravenna
three linked circles	descriptio	world	heirs

PLANS OF BUILDINGS

The 'Urbino plan'[27] was found on the Via Labicana and is now in the Ducal Palace at Urbino. It is a plan engraved on a stone slab of uncertain date, of an estate which includes a funeral monument, gardens, a plot for seed cultivation, a ditch, and an approach road, 1688 Roman feet long. The measurements of the property are given as 546 × 524½ Roman feet.

In the Palazzo dei Conservatori, Rome, is a mosaic plan of baths, also of uncertain date, found in Rome in 1872.[28] The figures on rooms of the bath-house do not, as was originally thought, refer to military units but to measurements of the rooms in Roman feet.

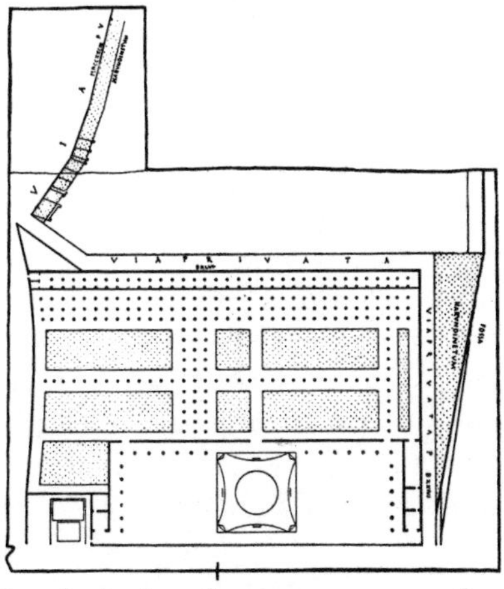

Fig. 19. The 'Urbino plan': plan of an estate on stone, found on the Via Labicana, now in the Ducal Palace, Urbino. Two *harundineta*, reed-beds, are shown, one near the public road, one near the private road with ditch marked beyond.

THE ARAUSIO CADASTERS

The three stone cadasters of Arausio (Orange) can be seen in the museum at Orange, except for some destroyed through a collapse of stonework in 1962, but available for study in the publication by A. Piganiol.[29] Cadaster implies a survey for taxation purposes, and we have a monumental inscription by Vespasian of AD 77 which may be translated: 'The emperor Vespasian, in the eighth year of his tribunician power, so as to restore the state lands which the emperor Augustus had given to soldiers of Legion II Gallica, but which for some years had been occupied by private individuals, ordered a survey map [*formam*] to be set up with a record on each "century" of the annual rental. This was carried out by . . . Ummidius Bassus, proconsul of the province of Gallia Narbonensis.' This refers to the Roman colony and territory set up at Orange about 35 BC. It was originally designed to provide land for veterans, and its first name was *Colonia Iulia firma Secundanorum*. Much later, perhaps in Domitian's principate, the name was changed to *Colonia Flavia Tricastinorum*, the Tricastini being the local tribe. This new name reflected a change in the apportionment of land, the Gauls acquiring many plots in the poorer areas. With regard to the abuse which was being rectified, to appropriate *ager publicus* had at least since the time of the Gracchi been an offence. In order to restore the depleted coffers, Vespasian reimposed revenues which for some years had not been collected, at the same time no doubt checking on the tribute due from the Gauls.

The stones were carved so as to show divisions into squares or rectangles, each representing one 'century'. In the case of the squares or nearly square rectangles of Cadasters B and C, the number of *iugera* in each 'century' was the commonest one of 200; whereas in the poorly preserved Cadaster A the highest number recorded is 330 *iugera*, so that each 'century' is likely to have contained 400 *iugera*. It is evident from aerial photography that these rectangular 'centuries' of double size were south of Orange on both sides of the Rhone. The tablets were housed in the local *tabularium*, record office, and according to one theory[30] Cadasters A-C had the top of the diagram facing south, west and north respectively and were attached to three internal walls of this building in such a way that viewers would be facing the

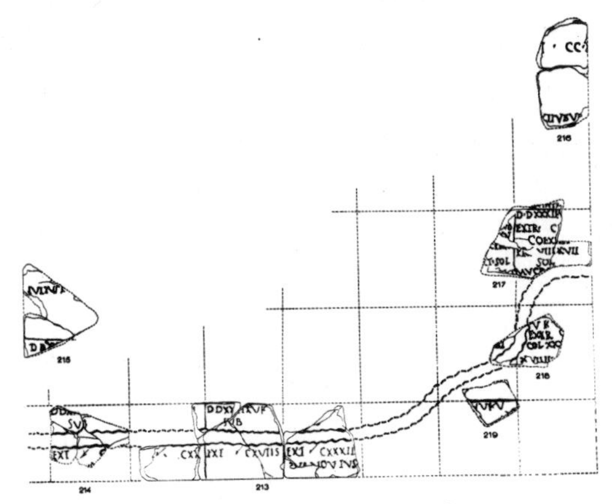

Fig. 20. Arausio (Orange), Cadaster B, Group IVF. Reconstruction of the position of almost adjacent tablets which show the course of a river.

appropriate direction. This theory however seems to require the inversion of B and C, since Cadaster B, which must have extended some 44 × 19 km in greatest measurements, was clearly north of Orange, towards Montélimar. We can certainly say that its *kardines* faced $5\frac{1}{2}^\circ$ east of north, but the exact location proposed by Piganiol is difficult to correlate in the area of the Donzère Gorge.[31]

The content of the squares or rectangles is as follows: (a) location co-ordinates with the abbreviations DD, SD, VK, CK explained on p. 90, followed by numerals; (b) status of land; (c) tariff and total rent payable; (d) details of rents. An interesting feature from the cartographic point of view is that rivers and pre-existing roads, at a different orientation from the centuriation *limites*, are entered. Cadaster C contains a large section in which islands in the Rhone called the *insulae Furianae* are shown, with centuriation and outline of islands; changes in the river bed at that point make it unlikely that these can ever be identified. In general we may say that there is enough cartographic content to label these tablets not as mere squares containing facts and figures but also as genuine thematic maps. According to correct procedure, in addition to a plan kept locally, a second copy should have been deposited in the *tabularium* on the slope of the Capitoline Hill in Rome (p. 96). No such cadastral maps, either from Orange or elsewhere, have been found there. One possibility for such total absence is that the copy deposited in the Rome *tabularium* may have been in bronze, which in the late barbarian invasions was seized and melted down. For surveyed land anywhere, Hyginus Gromaticus (p. 97) seems to envisage both copies of any map being etched in bronze, an extra piece being hammered on if necessary. The destruction of items on metal may explain why, although so many areas of centuriated land have been discovered and more await discovery, only one area, Orange, has yielded survey maps.

AQUEDUCT INSCRIPTIONS

Of Latin inscriptions relating to aqueducts, two are of special interest. The first is from Tusculum, of the first century BC, and illustrates with a small plan the authorization for certain landowners to tap the aqueduct water at particular hours.[32] One of the landowners is C. Iulius Caesar, presumably the dictator. A typical entry runs: *C. Iuli Hymeti Aufidiano aquae duae ab hora secunda ad horam sextam* (*Aufidiano* = 'on Aufidius' farm'). The method of drawing off water is illustrated in J. G. Landels, *Engineering in the Ancient World* (London, 1978), p. 50.

The other inscription is not accompanied by a map but reveals an interesting episode involving one.[33] About AD 150 Nonius Datus, an aqueduct surveyor, was asked to go to Saldae, Mauretania, where tunnel digging had resulted in failure. On his way he was stripped, beaten and robbed by brigands. When he arrived he found that the tunnels had been dug from north and south simultaneously, but both had veered to the right and had not met. He made a map for the procurator, new digging was based on this, and he was subsequently thanked for having solved the problem.

THE PESARO ANEMOSCOPE

An unusual Roman stone windrose, although only perhaps on the fringes of cartography, deserves a place in the history of Graeco-Roman maps. The 'Boscovich' anemoscope, in the Museo Oliveriano at Pesaro,[34] is a flat slab of Luna marble found on the Via Appia outside the Porta Capena, Rome, in 1759. Its inscription *Eutropius feci* indicates that it was made by a Greek, and it is thought to date to *c.* AD 200. In the centre is a circular hole which served to hold a pennant. Five parallel lines divide the upper

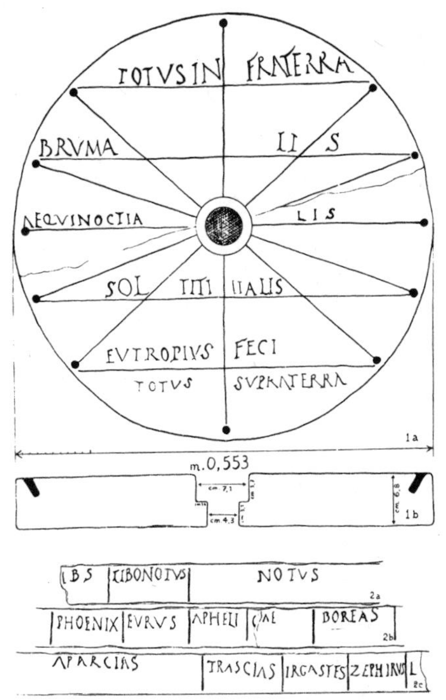

Fig. 21. The Pesaro anemoscope (see pl. 21). Wording on upper face: TOTVS INFRA TERRA(m); BRVMALIS; AEQVINOCTIALIS; SOL(S)TITIALIS (TI erased); EVTROPIVS FECI; TOTVS SVPRA TERRA(m). Names of winds on edge: LIBS; LIBONOTVS; NOTVS; PHOENIX; EVRVS; APHELI(otes); CAE(cias); BOREAS; APARCIAS; TRASCIAS; IRGASTES (= Ergastes); ZEPHIRVS.

surface into six zones. These lines are marked *totus infra terra(m)*, the Antarctic Circle; *brumalis*, the Tropic of Capricorn; *aequinoctialis*, the equator; *sol(s)titialis*, the Tropic of Cancer; and *totus supra terra(m)*, the Arctic Circle. The twelve points where these lines and a meridian drawn through the centre approach the circumference are marked by small holes which had bronze pegs; and in addition to the meridian, the other ten points are connected to their opposites by lines drawn through the centre. Round the edge are the names of the twelve winds, with unequally spaced dividing lines apportioning their central points to the twelve pegs. Clearly the anemoscope, with south at the top, was designed for travellers on the road south from Rome. It may be compared on the one hand with the windrose of Aristotle's *Meteorologica* (p. 28), on the other with the *Handy Tables* map (p. 170), which tries to combine an outline of the world with a windrose of this type.

CHAPTER VIII

ROAD MAPS AND LAND ITINERARIES

This chapter refers entirely to Roman times. There is no proof that the Greeks of the classical and Hellenistic periods had either road maps or land itineraries; though it is just possible that the lost map of Aristagoras (p. 23), which obviously depended to some extent on the Persian courier system, might be considered a forerunner.

The Roman military writer Vegetius, whose manual dates from after AD 383, uses the term *itinerarium pictum*, 'painted itinerary'.[1] By this he probably meant a road map, whereas an ordinary itinerary listed places on a road or on adjacent roads, with the mileage between each place. The need for road maps was no doubt particularly Roman, since their road network covered the whole Empire and was far superior to that of other ancient civilizations. Moreover the Roman habit of fixing milestones along all main roads (*viae consulares*) facilitated the development both of road maps and of land itineraries, since these gave some basis of scale, even though the maps were not drawn to a stated scale.

Into the category of such road maps only the Peutinger Table and the fragmentary Dura Europos shield fit. Ordinary land itineraries, on the other hand, complete or fragmentary, are not uncommon. The earliest of these date from the first century AD, whereas it has often been thought that road maps were not in use until the late Empire. But in this case we should not argue too much *ex silentio*. It will be seen that the Peutinger Table, which lists cities destroyed by Vesuvius, may have had a first-century AD predecessor. And it is wrong to assume that the text form must have preceded the graphic form, as shown by the fact that rudimentary cartography existed in many pre-literate peoples.[2] Instead it can be difficult to list places accurately without some pictorial basis, as can be seen, for example, from the confused lists of the Ravenna Cosmography (p. 174).

THE PEUTINGER TABLE

The Peutinger Table (Tabula Peutingeriana), now in the Nationalbibliothek, Vienna,[3] is a road map of the Roman world. Its name is derived from the scholar who owned it from 1508, Konrad Peutinger. It was in origin a long, narrow parchment roll, 6 m 75 × 34 cm, but has been divided into sections and placed under glass for purposes of preservation.[4] The existing map is a copy made in the twelfth or early thirteenth century, at several removes, from a road map of the fourth century AD. Two attributions can probably be dismissed: (a) that it was copied out by a monk of Colmar in 1265; (b) that it was originally Castorius' world map.

(a) On palaeographical grounds the extant copy of the Peutinger Table is almost certainly to be dated somewhat earlier than 1265. Also the number of sheets does not quite tally. The monk of Colmar copied his map on to twelve sheets. The Peutinger Table is on eleven: most of Britain, of Spain and of Morocco, originally on Sheet I, are not only missing now but were already missing at the time of the copying. This is indicated by the fact that truncated place-names have been given capitals as if they were whole names, e.g. Ridumo for (Mo)ridunum. So to make the number of sheets up to twelve, we should have to assume that there was an introductory sheet, which gave the title and other particulars, and that this was then detached and lost.

(b) K. Miller called the Peutinger Table 'the world map of Castorius',[5] but there is no good evidence for this. Castorius was a geographer of the fourth century AD frequently consulted by the Ravenna Cosmographer (p. 174). The latter's Cosmography has many names in common with the Peutinger Table; but it has far fewer names, and yet includes a number of places which are not on the Peutinger Table. Although there is a fair measure of correspondence in the areas where Castorius is mentioned as a source, there is also a fair measure of correspondence in Thrace, where the principal source is said to be one Livanius.[6] So it is safer to admit that we do not know who designed the original of the Peutinger Table.

The most obvious feature of the map is that its total width is about twenty times its height and was originally even more. Thus north–south distances, such as that between Rome and Carthage, look very short, while east–west distances seem much too long. Its object is to show main roads and the staging-posts on them, with distances between each. It is not to be thought of as typical of Graeco-Roman maps, which ever since Eratosthenes had, whether they contained co-ordinates or not, aimed at a reasonable representation of size based on latitude and longitude. There is no reason why road maps should have been thought of as needing special treatment in this respect.

The question therefore arises why there is such elongation. Normally the ancients thought of the inhabited world as having a proportion of 2:1 east–west/north–south. But Agrippa's map (p. 41), being affixed to a colonnade, may have been more elongated than this.[7] Moreover under the early Empire the normal medium for books, manuals etc. was the papyrus roll, which gave way to parchment for most purposes only about the third century AD.[8] The papyrus roll was wound on or back on wooden cylinders, and could be of considerable length,[9] though its width was limited. If the archetype of the Peutinger Table had a papyrus predecessor from which it was revised, its fourth-century revision will, while it changed the writing material, have retained the same format. The codex form, ancestor of our book, must have been considered less suitable in this case.

Treatment of towns on the Gulf of Naples may also suggest a first-century basis. Why otherwise should Pompeii, Herculaneum and the area round, destroyed by the eruption of AD 79 and thereafter little inhabited in antiquity,[10] appear on the Peutinger Table? Near by is a place called Oplontis, the *i* being restored (it appears also, as Opolontis or Eplontis, in the Ravenna Cosmography). Since 1964 a very large palace has been excavated at Torre Annunziata,[11] and the name has been rightly attached to it. If the Peutinger Table had been constructed *ab initio* in the fourth century AD, it would not have included such places as then no longer existed, many of which were presumably forgotten. There may have been an intermediate stage in the second or third century. The Antonine Itinerary could have been based on a map of that period. But to argue, with Kubitschek, that its wording implies the existence at one time of a world map commissioned by an emperor of the Antonine dynasty, seems unjustified.[12]

The map is not entirely confined to the Roman world: it includes certain areas to the north and east which lay outside the

Empire. Names of regions are given in capitals, but boundaries do not appear. Roads throughout are shown as a series of continuous lines, with numerals indicating distances between places. There is no difference between major and minor roads. The distances are normally in Roman miles (1 Roman mile = approx. 1.5 km); but in Gaul the measurement is in leagues (1 *leuga* = $1\frac{1}{2}$ Roman miles) and in former areas of the Persian Empire in parasangs (1 parasang = 3–4 Roman miles).

The Peutinger Table is clearly a civilian rather than a military map, since it has no military installations named. It has been thought by some to be connected with the *cursus publicus*, the postal service and system of staging-posts instituted by Augustus and intended mainly for those employed on public business and their entourage.[13] But there is no definite proof of this. The ordinary travelling public used the same roads and stayed at *mansiones*, inns (often in a town: see below), where they could if required change horses. For such travellers a road map would be a more useful guide than an itinerary, unless the latter was annotated and they were keeping to a specific route. Not much evidence either of date or of connection with the *cursus publicus* can be deduced from the inclusion of Aquae Labodes, Sicily, as a watering-place. This is known from an inscription to have been made between AD 340 and 352 into a station on the *cursus publicus*.[14] But whether the conventional sign would have changed from that of a spa if the map was made after AD 352 is uncertain.

Towns and other places are not marked by dots: the majority have no sign, though commonly a crook in the road, while a substantial minority have what the map-maker considered an appropriate conventional sign.[15] Of the 555 such signs on the extant part of the map, 429 show a façade with two towers, whether square or circular, probably intended to depict a villa;[16] 44 have a stylized building clearly intended to represent a temple, as shown by its application to Fanum Fortunae (Fano), where the Temple of Fortune gave its name to the town; 52, a surprisingly large number, are hollow squares or rectangles representing *aquae*, watering-places, especially spas. The remainder are a mixture including granaries, altars, *praetoria* (generals' headquarters or governors' residences), and lighthouses. In addition to the famous Pharos lighthouse at Alexandria, one on the Bosporus and one at Rome's harbour near Ostia are shown. One unique conventional

Fig. 22. The Peutinger Table: selected symbols for towns.

sign represents the *crypta Neapolitana*, a long, dark road tunnel between Naples and Pozzuoli.

Three large cities are given personifications:[17] Rome, Constantinople and Antioch. The goddess Rome is seated on a throne, wearing a crown (the original may have had a helmet) and red cloak. In her right hand is an orb, while her left holds a spear, and her shield is near by. This type of Rome personified has been traced back to Hadrian's reign.[18] The goddess is surrounded by a double circle, through which the Tiber flows, and round the circle are twelve main roads radiating out of Rome. The personification of Constantinople is similar to that of Rome, but there is no orb, and the headgear is a helmet. The right hand is pointing to what seems to be a column,[19] surmounted by a statue, perhaps the statue of Constantine which rested on a porphyry column. To these two cities, the obvious representatives of the Western and Eastern Empires, is added a third, Antioch. In the fourth century AD this played an important part as the city from which the defence of the Eastern Empire was organized. Its patron goddess, likewise seated on a throne, is of a type thought to be derived from a statue which Eutychides made in 296–293 BC of the Tyche (fortune) of Antioch. Nearby is the personification of a young man, who may be the god of the river Orontes. The temple to the left of these seems to be that of Apollo at Daphne, shown larger and more realistically than other temples on the map: the map-maker was presumably familiar with it. This temple was severely damaged by fire in AD 362, so that the original of the map may have been painted before that date. The arches below serve to represent the well-known aqueducts of Antioch.

An alternative explanation for the prominence of Antioch is to link the Peutinger Table with the map commissioned by Theodosius II, Emperor of the East AD 408–50 (p. 169). In his days Antioch was still important, but chiefly as an ecclesiastical centre. Certainly the Sinai area has two references to the Old Testament, which may be rendered 'the desert where the children of Israel under Moses wandered for 40 years' and 'this is where they received the law on Mount Sinai'. The New Testament is also involved, with a mention among other holy places of Mount Olivet. But the three personifications are at least in origin pagan rather than Christian, so that this theory can probably be dismissed.

In addition to these three, six cities are shown with fortification walls: Ravenna, Aquileia, Thessalonica, Nicaea, Nicomedia, Ancyra (Ankara: ancient name omitted). This type of representation can be paralleled by illustrations in the Vatican Virgil and in the Corpus Agrimensorum. Mapping is not confined to features connected with roads and towns. Large harbours are represented by semicircles. Mountains are shown as strips coloured pale brown, and important rivers are drawn as green lines.

The effect of the deformation is to stretch countries out on an east–west axis, so that the Levant coast, for example, is better seen if one allows for an orientation turned clockwise by about 80°. Although the considerable detail of the Nile delta comes out in roughly the right shape and orientation, the upper stretches of the Nile appear to be flowing from the area of the Atlas Mountains. This was a feature of early maps, such as that of Dicaearchus, but had been rectified by Eratosthenes. Carthage is nearer to Rome on the map than Sorrento is to Naples. Western Sicily looks as if it is adjacent to the Italian mainland near Salerno, while the south coast of Britain appears very close to the Garonne. Individual features, even in areas which should be familiar, are not infrequently out of proportion; thus Sardinia appears about eight times the size of Corsica, while many towns on the Adriatic coast of Italy are wrongly depicted as if inland. The remoter regions of the East are at times vague, and, as if to fill in empty spaces, descriptions appear such as *in his locis elephanti (scorpiones) nascuntur*, 'in these areas elephants (scorpions) are born'. But the area of the Persian Gulf contains a number of islands which have been able to be tentatively identified.

Britain was obviously shown with its main axis east–west. Only a fragment of the south survives,[20] the part which extended beyond the original first sheet of parchment. Whereas the main axis will have been lengthened, everything at right angles to it was shortened. There is an extraordinary telescoping of the south coast, since if we list, west from Kent, the places on or near the coast, we find (an asterisk in the following tables denotes a truncated name):

Peutinger Table	*Roman name*	*Modern name*
Ratupis	Rutupiae	Richborough
Dubris	Dubris (Dubrae?)	Dover
Duroauerus	Durovernum	Canterbury
Lemauio	Lemanis	Lympne
Iscadumnoniorū	Isca Dumnoniorum	Exeter
*Ridumo xv	Moridunum	near Sidmouth?

The only distance given in this section is between the last two named, 15 Roman miles. This tallies with the Antonine Itinerary; but the Moridunum in that itinerary (different from Moridunum = Carmarthen) is between Dorchester and Exeter. So this name

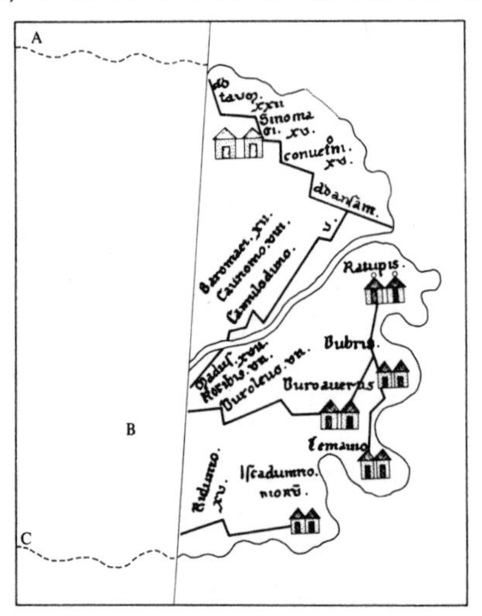

Fig. 23. The Peutinger Table: the British section, which is only partly preserved. A, C: conjectural continuation of coastline; B: approximate position of Londinium. The whole area from there to the north is lost (K. Miller attempted a reconstruction), and the existing south coast omits all coastal places between Lympne and Exeter and telescopes the distance between them.

ought to be 15 miles from Exeter on the way to Lympne. But that is only a minor error. The major error is the apparent omission of all intermediate places, which, coupled with a stretch of the lower Thames identifiable only as far as Crayford, near Dartford, makes it look as if Exeter and Moridunum are in Kent or Sussex. In fact one can imagine, as does K. Miller in his reconstruction, that the error is not one of omission but of the complete displacement of these two. Whichever is the explanation, no similar error occurs in other provinces of the Roman Empire. There was no coastal road all along the south coast; and as the compiler was evidently from the eastern Mediterranean, one may assume a lack of information about roads in Britain.

The stretch north of the Thames reads as follows:

Peutinger Table	*Roman name*	*Modern name*
*Baromaci xii	Caesaromagus	Chelmsford
Caunonio viii	Canonium	Kelvedon
Camuloduno v	Camulodunum	Colchester
adansam	Ad Ansam	Higham
Conuetoni xv	Combretonium	Baylham House, Coddenham
Sinomagi xv	Senomagus?	—
Ad *taum	Venta Icenorum	Caister St Edmund

Even in the eastern Mediterranean there are conspicuous mistakes; thus Gortyn(a) in Crete, which is called on the Peutinger Table Cortina, is made out to be a road centre near the northern coast of the island, whereas in fact it is on the Mesara, a plain of southern Crete. On the other hand there are examples also in this area of the Peutinger Table providing the first mention of places. Such a place is Bada in southern Macedonia.[21] This appears also in the Ravenna Cosmography, once as Bada and once as Bata. The entry on the Peutinger Table reads:

anamo vii
Bada xx
Arulos xv

meaning that Bada was 7 Roman miles from Anamo (Gannokhora?) and 20 Roman miles from Arulos, a misspelling of Aloros. Hammond, following Oberhummer,[22] equates it with Roman Pydna and takes it to be a corruption of the name Pydna;

while earlier scholars thought of it as a corruption of Balla. But surely it is more likely to be the Latin word *vada*, 'shallows', 'fords', in a spelling which reflects a Greek source; as there was no *v* in Greek, *b* was a frequent substitute. We learn from ancient accounts of the battle of Pydna[23] that the two rivers in the area were relatively shallow. We cannot be sure where Pydna was, though in ancient times it was on a hill near the sea. But the present writer would maintain that any Roman site near by was not also called Pydna but was called Vada or *Ad vada*, 'at the shallows'. For a Latin name cf. Arta below.

It should not be thought that the Peutinger Table was typical of Roman maps, or that the Romans were incapable of drawing a map to scale.[24] General maps, land surveyors' maps and town plans were drawn to scale, with reasonable approximations to cartography. If this road map was not, either it alone or the type constitutes an exception. Today's underground maps are not drawn to scale, but are really stylized topological models, simpler and more rapidly interpreted than a map to scale and in correct shape.

THE DURA EUROPOS SHIELD

In 1923 a fragment of parchment painted with a map was discovered at Dura Europos on the Euphrates (near Syria).[25] The finds were published and distributed between Paris, Syria and Yale

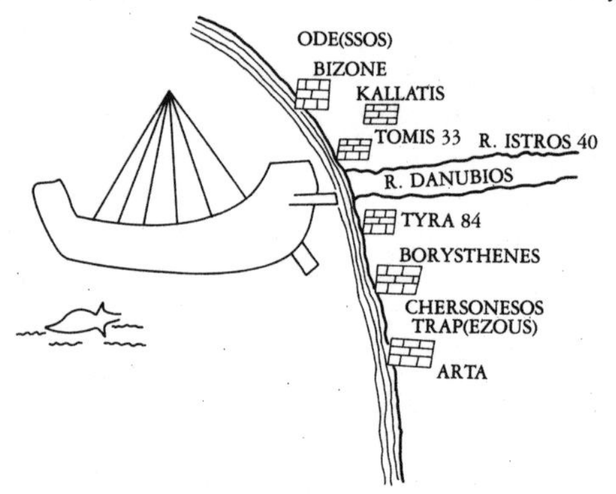

Fig. 24. The Dura Europos Shield, simplified redrawing of the north shore of the Black Sea, with Greek names Latinized. Trap(ezous) is not Trebizond but the Crimean table mountain, and *arta* may be a Latin word for 'straits' (of Kertsch).

University, and the map is now in Paris (Bibl. Nat., suppl. gr. 1354.5). It served to cover an infantryman's shield, and is datable to the years immediately preceding AD 260, when the Romans abandoned their fort. The extant portion measured, when found, 45 × 18 cm; the original length is thought to have been about 65 cm.

The Black Sea appears in blue on the left of the extant portion. It contains two ships, disproportionately large, and four men's heads, which may be those of sailors from other ships. The coast is represented by a pale curved line. On or near the coast are staging points stylized as houses with a few courses of pale green stonework. These and the principal rivers are given names in Greek, together with mileages (partly preserved) after the towns. The legible entries, with transcription of the Greek lettering, read:

Form given	*Mileage*	*Ancient name*	*Modern name*
Odess.	—	Odessos	Varna
Bybone	—	Bizone	—
Kallati.	—	Kallatis	Mangalia?
Tomea	33	Tomis	Constanța
R. Istros	40	Istros (town)	—
R. Danubis	—	Danubios	R. Danube
Tyra	84	Tyras	Akkerman
Bòrysthenes	—	Olbia on R. Borysthenes	Nikolaev
Chersonesos	—	Chersonesos Taurike	Crimea/Krim
Trap. . . .	—	Mt Trapezus (table mountain)	Krimskie Gory
Arta	—	*see below*	Straits of Kertsch

Although Istros or Histros was a name given to the lower Danube, the addition of *pot(amos)*, 'river', may be unjustified here, since there was a town of that name also and Danubis too is included.

The last two entries have been misunderstood. Trapezus is the correct expansion, but is not Trebizond, which is on the south coast of the Black Sea. It is the mountainous area of the Crimea, as pointed out by Uhden.[26] Then Arta is surely not part of Artaxata in Armenia, but is the Latin word *arta*, 'narrows', transliterated into Greek. The settlement near the Straits of Kertsch, to which this refers, was Pantikapaion. In the third century AD that town was ruled by a client prince but had a Roman garrison.

Thus the Dura Europos shield is far more correct in its cartography than used to be thought, and represents, fragmentary as it is, the nearest approach to the sort of military itinerary that Vegetius had in mind. It may well be unofficial, since the language is Greek and the inclusion of large ships on the Black Sea does not look like the mark of an official Roman army map. But the places and mileages will have been transcribed from such a map, which will necessarily have been in Latin. The road seems to have continued beyond the Straits. It was obviously of military importance, although no such road appears on the Peutinger Table or the Antonine Itinerary. It is, of course, possible that it was excluded from official maps for security reasons. But if no emperor of the Antonine dynasty visited the area, that would explain its omission in the Itinerary.

ROAD ITINERARIES

Organized Roman roads date from the early fourth century BC onwards, when the Via Appia was built from Rome to Capua and later on to Brindisi. Itineraries for the main roads in Italy may have existed from a comparatively early date, but we have none preserved earlier than the first century AD. The materials on which they were recorded are likely to have been mostly papyrus, parchment, stone or bronze. Extant itineraries are on parchment or, very fragmentary, on stone.[27]

The Vicarello Goblets

These most unusual itineraries, all of the same route, are on four silver goblets.[28] They were discovered at Aquae Apollinares (Vicarello, near Lake Bracciano), a thermal establishment, and are now in the Museo Nazionale delle Terme in Rome. From the name of the spa they are sometimes known as 'vases Apollinaires'. They vary from 9.5 to 15.3 cm in height, and each contains an itinerary from Gades (Cadiz) to Rome. Their date may be between 7 BC and AD 47 (possibly somewhat later), though they are not all contemporary. An indication of dating is given by the words *fines quadragesimae Galliarum*. This refers to a custom post for levying $2\frac{1}{2}\%$ customs duty. A starting but not an end date may be obtained from the wording *in Alpe Cottia*.[29] M. Iulius Cottius, a native prince, put up an arch to Augustus at Segusio (Susa) in 7–6 BC and

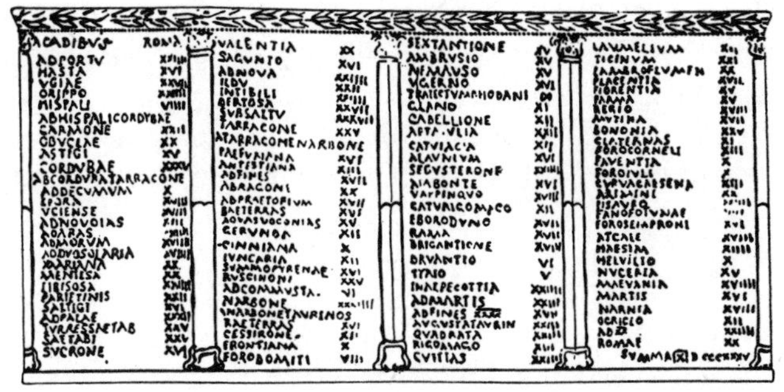

Fig. 25. Vicarello goblet: redrawing of the itinerary on one of the four silver goblets, 1st century AD, found at Vicarello on Lake Bracciano and now at the Museo delle Terme, Rome. Stages and distances in Roman miles from Gades (Cadiz) overland to Rome; cf. R. Chevallier, *Les voies romaines.*

improved the road over the Alps. Each of the vases has four columns of place-names and distances, these lists being separated by pilasters.

As with other Roman itineraries, each staging-point is followed by the number of miles from the preceding one. The route in Italy seems circuitous at first sight, but is in fact attested from other sources as a normal one. After Cordoba, Tarragona, Narbonne, Nîmes, Sisteron, and the Mont Genèvre, the traveller is taken to Rome via the Adriatic, a route designed to keep to low ground as much as possible. It therefore goes by Turin, Piacenza, Rimini, Fano and Bevagna; though it does not pass Vicarello, where the goblets were found. A Po valley section reads:

PLACENTIA	XVII	CLATERNAS	XI
FLORENTIA	XV	FORO CORNELI	XIII
PARMA	XV	FAVENTIA	X
REGIO	XVIII	FORO IVLI	X
MVTINA	XVII	CVRVA CAESENA	XIII
BONONIA	XXV	ARIMINI	XX

In this stretch between Piacenza and Rimini some of the figures are correct, some a little out, others corrupt; they may be compared with those of *itinera* on the Antonine Itinerary. Despite their comparatively early date, the Vicarello goblets display the same haphazard treatment of place-name endings as later itineraries. These are mostly in the ablative, but sometimes in the accusative, which seems the logical case, with a few in the locative.

It has reasonably been supposed that the owner, a Cadiz man, presented them to Apollo, the healing god, on a visit to Rome.[30] One would like to think, however, that in origin they had a utilitarian value. They could, for example, have been issued at various dates to reliable freedmen acting on behalf of a business firm as couriers or salesmen.

Isidore of Charax

Other extant land itineraries of the early Empire are in Latin, but the *Parthian Stations* of Isidore are in Greek, the lingua franca of the East.[31] His birthplace, Charax, was at or near the mouth of the Tigris (p. 50). About AD 25 he wrote a geographical work, of which the *Parthian Stations* may be an extract. It lists places from Zeugma on the Euphrates to Seleucia on the Tigris and Ecbatana and from there to Alexandria of the Arachosii (at or near Kandahar). A typical extract may be rendered: 'Next, the fort of Commisimbda on the R. Bilecha, 4 schoeni; after it Alagna, a royal station with fortified areas, 3 schoeni; after this, Ichnae, a Greek city founded by the Macedonians, sited on the R. Balicha, 3 schoeni.' The two variants for the same river show that the text is liable to corruption.

Hadrian's Wall Souvenir Itineraries

A series of forts on Hadrian's Wall is listed on three bronze vessels found respectively at Rudge, near Froxfield, Wiltshire;[32] at Amiens;[33] and (fragmentary with no text preserved) in north Spain. The list has been useful for identification of fort names, but there are no distances given, the names being accompanied only by a stylized castellated wall. The inscriptions are very similar, names running together in each.

Rudge: A. MAISABALLAVAVXELODVMCAMBOGLANSBANNA

Amiens: MAISABALLAVAVXELODVNVMCAMBOG. . .SBANNAESICA

Ancient name	*Modern name*
Maia (Maium)	Bowness-on-Solway
Aballava	Burgh-by-Sands
Uxelodunum	Stanwix
Camboglanna	Castlesteads
Banna	Birdoswald
Aesica (om. Rudge)	Great Chester

Whether these cups were intended as isolated souvenirs or as parts of dinner services is disputed. The list has helped because other sources are not quite conclusive on the identification of places in the western sector of the Wall and its extension.

The Antonine Itinerary

This is the nearest approach in list form to a road map, since it lists a number of routes, one often branching out from another. It is a manuscript itinerary consisting of a land and a sea section.[34] The titles of the two parts are

(i) *Itinerarium provinciarum Antonini Augusti*;
(ii) *Imperatoris Antonini Augusti itinerarium maritimum.*

It is obvious from the wording that they were drawn up in the first instance for land and sea journeys made by an emperor of the Antonine dynasty, AD 138–222 (except that Septimius Severus, 193–211, did not call himself Antoninus).

The longest journey is from Rome to Egypt overland via the Bosporus. This would, as has been pointed out, tally with Caracalla's journey to Egypt in 214–215;[35] his name was M. Aurelius Antoninus. Once the title had been applied to some of his journeys, it could easily be extended when journeys of later emperors needed to be provided for, and these could be tacked on without comment. Thus the capital of Macedonia, Pella, is presumably the place given as Diocletianopolis, assuming that about AD 290 its official name, after the Emperor Diocletian, was this. Either the collection may have been under active compilation for some seventy-five years or more, or old records may have been edited after 290 with updating of names.

Such routes were organized in connection with the *cursus publicus*, the transport and postal system established by Augustus chiefly for public employees and their families. This is not to say that the Antonine Itinerary consists of simple extracts from lists of the *cursus publicus*. Emperors' journeys normally had to be prepared well beforehand, with provision made at staging-points for food and accommodation for what was often a numerous retinue. Any particular journey by an emperor could be planned as a circuitous one, to take in places which needed to be visited. This is frequently reflected in the *itinera*, as will be seen in examples from Britain quoted below.

The *itinera* in Britain number fifteen, of varying lengths, with some overlap.[36] The traditional numbering, based on manuscript order, is as follows:

I From High Rochester to York and Praetorium (see below)
II From Birrens to York, Chester, London and Richborough
III From London to Dover
IV From London to Lympne
V From London to Caistor St Edmund, York and Carlisle
VI From London to Lincoln
VII From Regno (probably Chichester) to Winchester and London
VIII From York to London
IX From Caistor St Edmund to London
X From Ravenglass (Cumbria) to Manchester and Whitchurch (Shropshire)
XI From Caernarvon to Chester
XII From Carmarthen to Caerleon and Wroxeter
XIII From Caerleon to Gloucester and Silchester
XIV From Caerleon to Bath and Silchester
XV From Silchester to Dorchester and Exeter

As elsewhere, several of these journeys are very circuitous. Thus in *iter II* we may presume that after inspecting one or more military establishments, starting with Birrens in south-west Scotland, the Emperor needed to visit each of the two legionary fortresses which served as bases for the troops in the north, York and Chester, before returning to the Continent. This is the longest *iter*, said to total 481 miles, though the mileages given add up to 502.

Not all routes in Britain are certain throughout. Thus in *iter* I there are, admittedly, no problems except of measurement as far as York. But then we have entries reading:

Derventione	vii
Delgovicia	xiii
Pretorio	xxv

Derventio may be Malton rather than Stamford Bridge, but if so it is 20 Roman miles from York, not 7;[37] Delgovicia is unknown. *Praetorium* here means 'headquarters', but could be: (a) a corruption of Petuaria (Brough-on-Humber) or a reference to it; (b) a misplaced reference to York; (c) a place on the east coast, e.g.

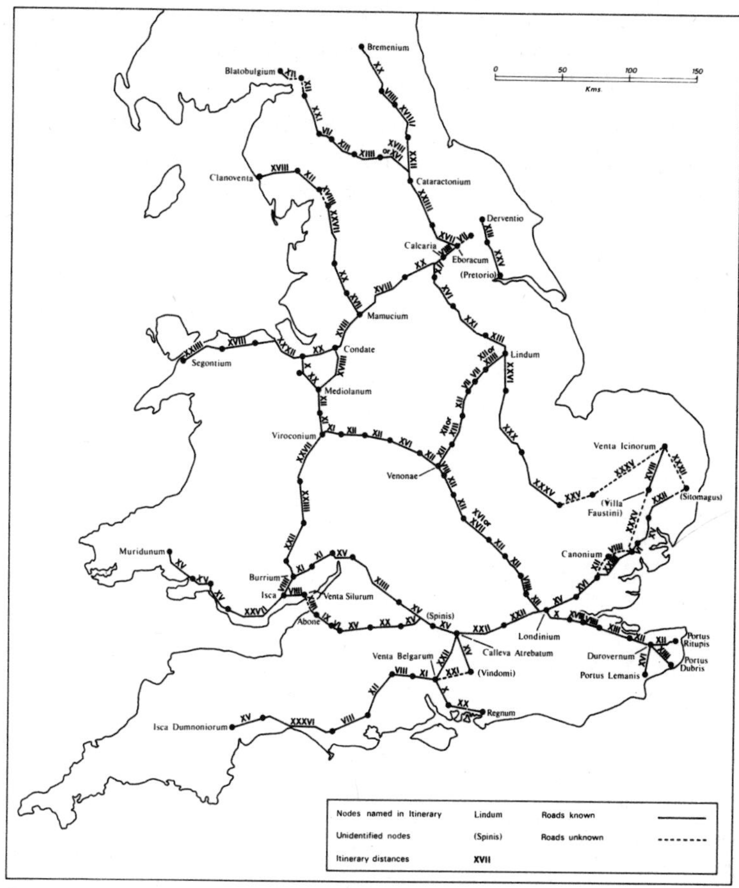

Fig. 26. Map based on the British section of the Antonine Itinerary, combining the 15 *itinera* and showing distances in Roman miles. Cf. Ordnance Survey, *Map of Roman Britain*, 4th edn.

Scarborough or Filey. If the route ends York-Malton-Brough, it forms a very acute angle.

The areas covered are all within the Roman Empire, though certain regions, such as the Peloponnese, Crete and Cyprus, have no entries, while others have only a few. The lists begin in Morocco, and some topographic arrangement is attempted, though it is not everywhere systematic. Each *iter* has its starting and ending points specified, together with the total mileage, which can be checked against the total obtained by adding the stages. These too, however, not infrequently have corrupt numerals; and where a section of road is repeated between one *iter* and another, the distances do not necessarily agree.

Despite its defects, the Antonine Itinerary is an extremely valuable document for tracing, alongside the Peutinger Table and other sources, the numerous staging-points on the network of Roman roads. Maps can be reconstructed from the lists, and for many areas this has been done.[38] A study of mileages based on known sites has established that from a large town the distance was sometimes calculated from the centre, sometimes from the outskirts.[39]

The Bordeaux-Jerusalem Itinerary

Itineraries were constructed not only for official purposes and for the casual traveller, but also, as Christianity grew, for pilgrimages. The aim of these was the Holy Land, and from the time of Constantine the Great many such pilgrimages were made. For most of the more distant pilgrims, the further stretches of the route were unfamiliar. Moreover there might be important shrines or relics to visit, either on the way or a short distance from it.

The best specimen of such an itinerary, usually known as the *Itinerarium Burdigalense*, dates from AD 333.[40] It starts at Bordeaux, a great cultural centre, and first makes for Narbonne. Then it goes via Arles, Valence, Gap, Susa and Turin to Milan, where the alternative route, as will be seen, turned off towards the south of Italy. The main route, however, is given as the outward one. This takes the pilgrim past Aquileia, Poetovio (Ptuj), Sirmium (Sremska Mitrovica), Naissus (Niš), Serdica (Sofia) and Constantinople, where the Bosporus is crossed. Here a personal note is introduced. 'We walked from Chalcedon (to Jerusalem) starting on May 30 in the consulship of Dalmaticus and Zenophilus (AD 333) and returned to Constantinople on December 26 of the same year.' The route of the anonymous pilgrims took them via Ancyra (Ankara), Tarsus, 'from which the apostle Paul came', Tyre and Caesarea. 'There is there', they write, 'a bath of the centurion Cornelius, who gave out many alms. At the third milestone is Mt Syna, where there is a spring in which if a woman washes she becomes pregnant.' At Neapolis (Nablus) the regular mileages cease and a guided tour of the Holy Land begins.

For the return journey, after Heraclea in Macedonia an

alternative route is given via Aulon (Vlorë, Albania), Brindisi, Capua, Rome, Fanum Fortunae (Fano) and Milan. Presumably the pilgrims would want in one direction to see the holy places of Rome, otherwise they would have had a much more direct journey along the Adriatic coast. This itinerary marks the end of a series which had a basis of accurate measurement; others do not give the distances, in fact often have little to say on any areas except the Holy Land itself.

The chief users both of road maps and of itineraries were army personnel and civilian travellers. There were many long-distance travellers on foot, and the medical writer Galen recommends this for keeping fit. There was also a significant amount of tourists and of visits to seaside resorts and spas. It seems almost certain that there was little demand from carters. The cost of land transport was so high, as compared with sea transport, that it tended to operate locally on well-known routes.

These *itineraria* very reasonably listed all possible stopping-places, and we should be wrong in thinking that a traveller with or without a vehicle would necessarily think of stopping at the next staging-post. The mileages covered by Horace's party on their way from Rome to Brindisi,[41] where ascertainable, average out at 18 Roman miles a day. So the mere fact that the section of the Bordeaux–Jerusalem itinerary between Arles and Milan, including the Alpine tract, has stopping-places with an average distance of only six miles[42] is not necessarily significant. But as the original party took nearly seven months to walk to the Holy Land from the Bosporus, see its sights thoroughly and return, we can see that they were in no hurry.

CHAPTER IX

PERIPLOI

The word *periplus* ('a sailing round') can mean a circumnavigation or other coastal voyage.[1] From the point of view of contribution towards map-making, however, only such circumnavigations or other coastal voyages as were recorded are of major interest; these records too were called *periploi*. In the Mediterranean world of antiquity the Phoenicians, with their colonists the Carthaginians, and the Greeks were the most famous navigators. The voyage sponsored about 600 BC by Pharaoh Neco (p. 133) was manned by Phoenicians, who reported that on their clockwise circumnavigation of Africa the sun was on their right. This may have helped to convince some of the ancients that Africa extended into the southern hemisphere. Only one Carthaginian *periplus*, that of Hanno, has survived, and that in Greek translation, though we are able from Greek geographers to gain some impression of other voyages. This paucity is perhaps due either to the comparative lack of Phoenician and Carthaginian material, a lack paralleled in the literary field, or to an unwillingness to translate into Greek or Latin any writings unless they offered special information or could be regarded as major works of literature.[2]

Coastal voyaging and island-hopping were the normal method in the areas navigated by the Greeks and Romans, especially in the winter, which was regarded as a season when only profit-conscious ships' masters were willing to venture out of sight of land. The Romans preferred to travel long distances by land rather than sea even when a much longer journey was involved. Whereas land journeys could be measured fairly accurately, early measurement of sea journeys tended to be merely days of sailing. In some cases we are told that this was following coastal indentations, in some from point to point; in some neither is specified. Although winds and currents caused considerable variation in mileage covered, it was generally reckoned that a day's sail averaged about 56 Roman miles.

THE EARLY PERIOD

Although Homer's *Odyssey* could be thought of as based on a *periplus*, it does not set out to be a geographical poem, as, in the sphere of much later epic, Apollonius Rhodius' *Argonautica* (third century BC) may be said to be. The places which feature in Odysseus' wanderings outside the Greek world cannot be conclusively identified,[3] though some were in Classical antiquity (from a comparatively early period, but with increasing impact in Alexandrian times) given plausible equivalents, such as the Island of Djerba, Tunisia, for the Land of the Lotus-eaters. When Homer speaks of a sea voyage lasting nine days,[4] he is using a poetically significant numeral (there were for example nine Muses), not one associated with exact calculation. It has been conjectured that Homer's Cimmerians, living at the edge of the Ocean in a dark fog, and his Laestrygones, giants living where the paths of day and night are close to each other, represent races bordering the north Atlantic.[5]

Among early Classical *periploi* of which we hear at second hand are those written about 550–525 BC by two ships' captains from Massalia (Marseilles), colonized by Greeks from Phocaea in Asia Minor. The first, Euthymenes, explored the west African coast perhaps to Senegal or The Gambia,[6] and claimed that a branch of the Nile rose near that coast. With regard to the second it is said, according to verses of the late Roman writer Avienius, that ships from Tartessus regularly went to Oestrymnis, equated by modern scholars either with western Spain or with Brittany.[7] Tartessus was the Tarshish of the Old Testament, and may represent not a town but a coastal zone between Gibraltar and Cape St Vincent. The timing of two days' sail from Spain or Brittany to the sacred island inhabited by Hibernians (inhabitants of Ireland) seems insufficient, and Britain, as recorded in Avienius, is barely mentioned. It seems likely that the original was a versified Greek account, perhaps likewise imprecise and thus creating a double error. Nevertheless, such information was of use to later map-makers; it was their source, since for a long period from about 500 BC the Phoenicians and Carthaginians took care to exclude other maritime nations of the Mediterranean from penetrating, for either mercantile reasons or scientific exploration, beyond the Straits of Gibraltar.

Some Carthaginian explorations, however, were translated from Punic into Greek. One such expedition was that of Himilco of Carthage from the Straits of Gibraltar to Brittany, but little is known of this. The account of which most has come down to us is the *periplus* of Hanno the Carthaginian.[8] It is thought to have taken place before 480 BC, but the Greek translation is later. The text begins: 'The Carthaginians resolved that Hanno should sail beyond the Columns of Hercules [Straits of Gibraltar] and found cities for settlement of Libyphoenicians [inhabitants of the Carthaginian hinterland]. He therefore sailed with sixty penteconters (fifty-oared ships) and a total of up to thirty thousand men and women, with corn and other provisions. Two days' sail beyond the Pillars we founded the first city, which we called Thymiaterion; it had a large territory.' The reason for the move was no doubt over-population. Thymiaterion can be identified with Mehedia in Morocco, and the translator has given the Punic Dumathir a similar-sounding Greek name, ?'censer'. Some others of the colonies founded can likewise be identified, such as Akra (Agadir). But as the account progresses, perhaps because of intentional omissions in the Punic original,[9] together with the unfamiliarity of the terrain in antiquity, the localization of places mentioned, despite indications of distance, becomes very difficult; though the high wooded mountains described as being twelve days' sail south of Cerne island seem to be Cape Verde. The most conspicuous natural feature is a mountain which they called, in the Greek translation, Theōn Ochēma, 'Chariot of the Gods'. This is said to have been a very high active volcano in a volcanic area, and is generally identified with Mt Kakulima in Guinea. The account continues: 'Sailing for three days from here past rivers of lava we reached a bay called Notu Keras ["Horn of the South", identified with Sherbro Sound]. In a recess of this bay is an island like the first one, with a harbour. And in the harbour was a second island, full of wild human beings. Most of these were women with hairs on their bodies, whom the interpreters called Gorillas [perhaps orang-utans]. We chased the men but could not catch them, as they were agile on the rocks and defended themselves with stones. But we captured three women, who bit and scratched their captors and were unwilling to come. We killed and flayed them and took their skins to Carthage. For we did not sail any further, as our food was running short.'[10]

It seems unlikely that Hanno had read the account of the expedition sent round Africa by Pharaoh Neco, though in any case he may not have wished, like its sailors, to wait for the sowing and reaping of crops. Hanno's account was known in Roman times, though the elder Pliny in one passage (p. 69) implies that the work was considered lost. Later he gives a second-hand account,[11] differing in that (a) the island is said to be near Cape Hesperu Keras ('Horn of the West'), whereas Hanno says it was on the bay called Notu Keras; (b) the islands are said to be inhabited by Gorgons; (c) they are said, according to Xenophon of Lampsacus, to be two days' sail from the mainland; (d) Hanno is said by Pliny to have placed skins of two, not three, dead Gorgon women in the temple of Juno (= Tanit) at Carthage, where they remained until its capture by Rome. Sallmann thinks this account of Pliny's may come from Cornelius Nepos, whom Pliny elsewhere considers unreliable.[12]

Whereas Hanno's voyage was primarily for the purpose of colonization, we do hear through Herodotus not only of the earlier voyage sponsored by the Pharaoh, but of one Sataspes, a member of the Persian royal family, who was sent to sail round Africa from the Straits of Gibraltar to the Red Sea or Persian Gulf, but twice turned back.[13] He seems to have taken only one ship, and to have reached further south than Hanno, reporting on the local tribes; but he was unfortunate enough to be executed by Xerxes on his second return.

The question how these early sea voyagers found their way is difficult to answer. There is no evidence that they followed drawn charts, yet already the concept of a continent which could be rounded had been formed. The ancients had different views about areas of greatest heat, and many disbelieved in an inhabitable southern hemisphere, but in the early period all thought of some sort of encircling ocean, a concept which encouraged attempts at circumnavigation.

PS.-SCYLAX AND THE HELLENISTIC AGE

It has been shown from an examination of the status of Greek cities that the *periplus* of ps.-Scylax is not earlier than the fourth century BC and is probably to be dated between 361 and 357 BC. It was therefore not written in the extant form by the well-known Scylax, a Carian who between 519 and 512 became admiral of the Persian fleet under Darius I. This admiral sailed down the Indus, starting from Caspapyrus or Caspatyrus, on the Kabul river or at its confluence with the Indus, to the mouth of the Indus and eventually after two-and-a-half years to Arsinoe near Suez. Seven fragments are thought to come from his account of this navigation. The *periplus* of ps.-Scylax[14] is a compilation from earlier *periploi*[15] and from accounts of the fourth century BC. An extract from it, referring to part of the Tunisian coast from the Libyan frontier westwards and northwards, reads in translation: 'After Abrotonum the city and harbour of Tarichiae; the sail takes one day. Seaward from Tarichiae is an island called Brachion[16] [later Meninx, now Djerba] or island of the Lotus-eaters. This is 300 stades [= 36 Roman miles[17]] long, somewhat less in width. It is about 3 stades from the mainland. On this island grow the edible lotus and another from which they make wine. The fruit of the lotus is as big as that of the arbutus [strawberry-tree]. They also make much oil from wild olives. The island has a good soil and grows good wheat and barley. It is one day's sail from Tarichiae to the island. After it is the city of Epichos [probably corrupted for Gichtis or Gigthis; ruins on the shore at Djorf bu-Ghara]. From the island to Epichos a half day's sail.'

It must be admitted that this quotation is not quite typical of the entries, since the section relating to Djerba is given in far greater detail than many other entries. With its probable Homeric association and its contemporary (fifth/fourth-century BC) prosperity, it clearly appealed to the writer, who may well have stayed some time on the island, at that time more fertile and productive than the new Arab landscape which meets today's tourists. One may call the work a combination of what is now known as a 'verbal map' with a selective guidebook. Such a verbal map, with or without this addition, would provide much of the material for the compilation of a coastal map.

Just as Alexander the Great chose experienced surveyors to measure and prepare descriptions in advance of land to be traversed, so he chose Nearchus the Cretan, an experienced ship's captain, to command his naval expedition in the East and write up an account of this.[18] A fair amount of this report is known to us through Arrian's *Indica*. When Alexander's soldiers refused to march further east, the King planned a combined land and sea expedition down the rivers Hydaspes, Acesines and Indus and along the coast to the Persian Gulf. This section of the Indian Ocean had been explored by Scylax, but he was relatively little known by Greeks of Alexander's time. 'Alexander had a desire', says Arrian, 'to sail right round [he presumably means only round the coast of] the Indian Ocean from India to Persia, but he was worried at the length of this voyage and afraid that they might encounter a place with no harbours or an appalling climate, and that the fleet would be lost.'[19] However, Nearchus obviously persuaded him that the expedition was feasible, and was told to record coastal features, tribes etc. He wrote in discursive, Herodotean style, and started recording animals in India, but had to admit he had only seen a tiger skin, not a live tiger. Unfortunately Nearchus' fleet was held back at the Indus delta by adverse winds. Meanwhile Alexander, though 'not ignorant of the difficulties',[20] kept close to the shore of the Makran, with its terrible heat, sandhills and snakes. It was not till two months after the army set out that the fleet was able to make good progress, so that it could not bring provisions to the army. Nearchus made notes of the voyage, including ports where spices like cinnamon arrived from Malaysia, and particulars of great whales encountered in the Indian Ocean. The places mentioned are not identifiable, and there is a conflict between Arrian, who quotes measurements in stades, and Pliny, who says that the report has no measurements or staging-posts.[21] One claim can certainly be discounted: according to Nearchus, he had been so far south that the sun was north of him, i.e. that he had crossed the equator, whereas what is known of the expedition shows that it followed Alexander's advice to keep near the Baluchistan shore, finally meeting him at the Straits of Hormuz.[22]

The *periplus* of the Persian Gulf was written up in the lost work of Androsthenes of Thasos.[23] Much of the information preserved on this is in Theophrastus and concerns the flora and fauna. But we are told by Strabo that Androsthenes reckoned the Arabian coast as 10,000 stades long, that he made the sea voyage by himself as well as with Nearchus, and that he was interested in signs of Phoenician-type temples and an island called Tyros (Tyre).

This happened in the late summer and autumn of 325 BC. At much the same time at the opposite extremity of the known world another Greek-speaking navigator was recording two *periploi*. The expeditions of Pytheas of Marseilles are thought to have taken place about 320 BC.[24] His accounts, whose Greek titles may be translated *On the Ocean* and *Periplus* or *Journey round the world* (*Gēs Periodos*), have not survived, but we can reconstruct a fair amount from later writers. Of these, Eratosthenes and Hipparchus reported favourably, but Polybius and Strabo criticized what they had read of his voyages and considered him an arrant liar. It seems likely that he made two voyages rather than one: (a) both coasts of Britain, together with Thule; (b) 'the whole coastline of [western and northern] Europe from Gades [Cadiz] to the R. Tanais [Don]'.[25] Since Strabo summarizes the two as 'the whole of northern Europe', the wording for the second voyage cannot refer to the Mediterranean; presumably he thought the Don flowed into a northern sea such as the Baltic. Pytheas' estimate for the circumnavigation of Britain, 40,000 stades (with the commonest contemporary equivalence, 4800 Roman miles), is wildly exaggerated. If, however, he followed all the principal coastal indentations from Land's End to the north of Scotland, and reckoned his journey in terms of days' sail, it becomes somewhat more comprehensible. The northernmost point reached was the island of Thule,[26] which according to Pytheas, paraphrased by Pliny,[27] was six days' sail northwards from Britain and had continuous daylight in summer, continuous night in winter. The Thule seen by Agricola's fleet[28] when it was cruising near the Orcades (Orkney) was clearly Shetland, but this does not prove that Pytheas' Thule was likewise. Scholars have tended to favour part of the Norwegian coast mistaken as an island, though Hawkes argues for Iceland.[29] Both of these would conform to the length of daylight. But if one sails north from Cape Wrath for about 300–350 Roman miles, one reaches the Faeroes, through which the 62° N. parallel runs. Admittedly this is not inside the Arctic Circle, but neither is most of Iceland; and Norway, like Iceland, is much further than Pliny intimates. In Ptolemy's *Almagest* the island of Thule is given as 63° N., with a longest day of twenty equatorial hours.[30] Since we know that at several places Pytheas made observations of the sun's angle at midsummer, his account of Thule may have included one such, which will have misled such later writers as took Thule to be Shetland, whereas it may have referred to the Faeroes.

If on the second voyage Pytheas set out to travel as far as the Tanais (Don), probably his idea was that the Don could in some way be reached from the Baltic. It has long been suggested that the city of Massilia was interested in the lands from which amber came. The latitude corresponding to 54°14′ which he gave to an island called by him Abalos, perhaps 'apple island', and by Timaeus Basilia, Kingly, has been thought to refer to the northern part of Jutland.[31] His estimate of the length of the sea penetrated by him, 6000 stades, suggests that when he reached the point where the Baltic broadens out, he thought he had reached the Northern Ocean and turned back. The main amber routes led from the Baltic to the Black Sea, and he may have wanted to trace one as far as the R. Don.

Although Pytheas was dismissed as a teller of sea yarns by some later writers, he seems to have been more methodical than many ancient navigators; the cartography of Eratosthenes (p. 33) was obviously indebted to him and his findings. The northern waters were relatively uncharted, and Greek cartographers of the Hellenistic Age turned to Pytheas as one who at least observed, measured and recorded latitudes and distances.

Coastal exploration under the Ptolemies of Egypt was designed to promote trade in southern latitudes. Piracy in the Red Sea was suppressed, agreements were made with the Arabs of Yemen, and trading stations were set up along the east coast of Africa. We do not know whether the area was mapped; though if not, advances were certainly made in the plotting of coastal distances and features. But one of the areas developed was the Cinnamon country, the Somali coast west of Cape Guardafui, at a latitude of 11°–12° N. Whereas in later cartography more northerly areas of Africa were given parallels of latitude based on places on the Nile, not necessarily all on the same longitude, the Cinnamon country served as the parallel for this southerly zone.

ROMAN SEA VOYAGES IN THE NORTH-WEST

In spite of greatly increased trade and travel with Rome's expansion, there was no great increase in compilation of *periploi*. The Romans were not in general lovers of sea expeditions, but the more commercially minded among them did sponsor Greek ships' captains to develop trade. Strabo shows the extent of the rivalry between Rome and Carthage when he recounts how the captain of a Carthaginian ship, making for the ten Cassiterides (Tin Islands) from the Spanish coast to load tin, was publicly rewarded when he grounded his vessel on a sandbank rather than continue and so indicate his route to the following Roman ship's crew.[32] He goes on to mention a Roman, P. Crassus,[33] who opened up the approach to the Cassiterides, which he calls a voyage 'further than to Britain'. No conclusive proofs have been adduced to identify these islands with Cornwall[34] or other areas of the British Isles. Presumably P. Crassus and his successors who made this voyage were equipped with either a *periplus* or a rudimentary map.

THE ERYTHRAEAN SEA

The *Periplus of the Erythraean Sea*[35] is a Greek prose work of the early Roman Empire. Two kings in different areas mentioned in the text are Malichus (the name means 'king') the Nabataean and Mambanos of the Bombay region. Recent scholarship is inclined on balance to give a first-century AD date for these two kings.[36] Between AD 40 and 70 Malichus II was ruler of the Nabataeans, whose capital was Petra. Mambanos may or may not be identical with Nahapāna, a king of western India whose date is controversial: earlier he was thought to have died about AD 125 or 130, but now *c.* AD 100 is thought perhaps a more likely date; if so, he could, if he had a very long reign, have started it as early as the end of Malichus II's reign. The unknown author was a trader, who describes two sea voyages which started down the Red Sea,[37] the first to the Far East and the second to east Africa. It was not the same style of work as the treatise *On the Erythraean Sea*,[38] written about 132 BC by Agatharchides, an untrustworthy geographer. Although the writer of the *periplus* is familiar with the Indian coast as far as the Ganges, he does not know Sri Lanka. An extract may be rendered: 'This river [the Sinthos, i.e. Indus] has seven mouths, but they are shallow and marshy, so that they are not navigable except for the central channel, on which there is also a sea trading post, Barbarikon [Bahardipur]. Opposite it is a small island, and in its hinterland the capital of Scythia, Minnagar. It is ruled over by factions of Parthians, who are constantly ousting each other.'[39] The part played by Mediterranean traders in this area is attested by coins and Indian sources. Beyond India the account becomes sketchy and ends at Thinae, an inland city from which silk and Chinese cloth was taken.

The chief difference between this *periplus* and others lies in its concentration on trade. The number of items traded in is very large; thus those listed for Barugaza (Broach, India), as recorded in § 49, are:

IMPORTS	EXPORTS
(a) *For the market*	Spikenard, Saussurea Lappa (root used as a spice), balsam gum
Wine, chiefly Italian, some from Laodicea and Arabia	Ivory
Copper, tin, lead	Onyx stones, myrtle- (or mulberry-) coloured stones
Coral, topaz	Lykion (an Indian plant)
Clothing, unlined and inferior, of all kinds; damask girdles 1 ft long	Linen cloth of all sorts, silk, mallow-cloth, yarn
Storax gum (Styrax officinalis)	Long pepper
Sweet clover	Goods from other markets
Unworked glass	
Red sulphate of arsenic (realgar); powdered antimony (kohl)	
Gold and silver coinage, which can be exchanged at a good rate for local coinage	
Perfume, but not expensive nor in large quantities	
(b) *For the king*	
Expensive silver plate	
Music	
Attractive girls as concubines	
Quality wine	
Expensive unlined clothing	
Choice perfume	

The section ends: 'Sailings from Egypt to Barugaza about July'.

The *periplus* is less detailed about the east African coast. Menuthias is described[40] as 'an island about 300 stades from the mainland, low and wooded, in which there are also rivers and many types of birds and a mountain tortoise. It has no wild animals at all except crocodiles, which do not harm anyone. It also has boats that are sewn together (*rhapta*) and made of a single trunk;[41] these they use for fishing and catching turtles. . .'.[42] Although the distance from the mainland, equivalent to 37½ Roman miles, would suit Pemba better, mention of rivers seems to point to Zanzibar. Two days' sail beyond Menuthias was 'the Rhapta', named after the same type of boat as was mentioned before. This settlement is described as selling ivory and tortoiseshell, and as having a chief dependent on Arabs from Mouza (Maushij). In Ptolemy the latitude of Rhapta is given as 7° S.; this suits Dar es Salaam, to which two days' sail sounds too little from Zanzibar and too much from Pemba, or the Refiqi delta, rather more than two days' sail from Zanzibar. The name of the river Refiqi, contrary to the usual tendency, might have been derived from the town Rhapta.

ARRIAN

The work of Arrian[43] (Flavius Arrianus) shows how a *periplus* could be of military use to the Roman Empire. Hadrian made it clear that he welcomed all military information of interest from provincial governors; and Arrian's *Periplus of the Euxine* (Black Sea) is addressed to that emperor in the form of a letter. A Greek from Bithynia, he had become governor of Cappadocia under Hadrian and had defeated an invasion by the Alani in AD 134. An extract from the *periplus* may be rendered: 'Before midday, having sailed over 500 stades [from a village on the Black Sea called Athenai], we reached Apsarus, where the five cohorts are stationed. I gave the soldiers their pay and inspected their arms, the wall, the ditch, the sick man and the existing state of provisions. My opinion about this has already been expressed in my Latin letter. It is said that Apsarus is the same place as was of old called Apsyrtus, named after Medea's brother Apsyrtus who was killed there . . .'.[44] It appears from the previous context that Arrian's ship had to tack on leaving the village, so that the recorded distance, which depended on hours of sailing time, is more than would have been expected.

OTHER PERIPLOI OF THE LATE EMPIRE

Substantial fragments of the *Stadiasmus Maris Magni*,[45] a Greek *periplus* of the Mediterranean of *c.* AD 250–300, survive: (a) from Alexandria along the north African coast to Utica, (b) from Phoenicia to Crete. The entries are mostly only of distances, which in the case of well-known Greek islands radiate out; but occasionally there are descriptions such as the following: 'As you sail in, you will see low-lying land with islands. When you get near them, you will see the city near the sea and a white sandy beach; and the whole city is white. It has no harbour; but you can moor safely near the temple of Hermes. This city is called Leptis [Lepcis Magna].'[46]

In addition to translating Dionysius Periegetes (p. 143), the Roman senator Rufus Festus Avienius (fourth century AD) wrote in Latin verse a work called *Ora Maritima*,[47] 'sea-shore'. In its extant form it is brief and thought to be incomplete. From Massilia (Marseilles) it gives a description of the coasts as far as Gades (Cadiz), together with an account of maritime exploration from Cadiz, including that of Himilco (p. 132). As the coastline west of Marseilles has in places changed appreciably since ancient times, it is of interest to have Avienius' account of it, and during the present century a cartographic reconstruction of it has been made.[48]

By this time Roman forces and Greek and Roman traders had penetrated many areas previously not well known. Marcian of Heraclea Pontica (*c.* AD 400) is the author of a Greek *Periplus of the Outer Sea* in two books;[49] the description of some areas is fragmentary or missing. The 'outer sea' is what in earlier times was called the Ocean; and Marcian's work gives us something more than mere description. He refers among his sources to the *Geography* of 'the most divine and wise Ptolemy'.[50] Book I deals with the East and Book II with the West. The author shows a more scientific attitude to measurement than those of other *periploi*. Distances are given in stades: for Europe two figures are recorded, indicating maximum and minimum mileages; for other areas only one figure is given. Previous writers, he says, have tended to give only one figure throughout, as if distances traversed by sea could be measured with a rope. He also tells us in his preface that he will include the major islands, 'the one called Taprobane, formerly called the island of Palaesimundu[51] [Sri Lanka] and the two Britains' (Britain and Ireland). The entry on Britain shows an interest in statistical analysis. Marcian's paragraph (ii.45) may be rendered: 'The length of the British island of Albion[52] begins at the western horizon at the Damnonian promontory, also called Okrion; it ends at Tarvedunum, also called the promontory of Orcas; so that its greatest length is 5225 stades (= $653\frac{1}{8}$ Roman miles). Its width begins at the Damnonian promontory, also called Okrion, and ends at the peninsula of the Novantai and the promontory of the same name, so that its greatest width is 3083 stades (= $385\frac{3}{8}$ Roman miles). It embraces 33 tribes, 59 well-known cities, 40 well-known rivers, 14 well-known promontories, 1 well-known peninsula, 5 well-known bays and 3 well-known harbours. The *periplus* of the whole island of Albion is not more than 28,604 or less than 20,526 stades' (= $3575\frac{1}{2}$–$2565\frac{3}{4}$ Roman miles). Similar statistics are available for a number of other regions. Can we deduce from this that not only with regard to Britain, but with regard to these others, it is apparent that Marcian had a map or maps to hand? Since he mentions Ptolemy but not Marinus, it may be that he had some sort of Ptolemy map, but perhaps only a world map, such as Agathodaimon composed. Ptolemy does not mention distances within Britain except in criticism of Marinus. But Marcian can have found his measurements in his other source, one Protagoras (p. 155). The odd shape of Britain ascertainable from Ptolemy's co-ordinates, no doubt influenced by Marinus, and visible in Ptolemaic maps, has strongly influenced Marcian. His length is SW–NE, from The Lizard to Dunnet Head. His width is south–north, from The Lizard to the Rhinns and Mull of Galloway, wrongly thought of by Ptolemy as the northernmost point of Britain. The measurements are very strange. On the one hand the length and width are far too short, since if a regional map is drawn based on Ptolemy's co-ordinates and with his recommended proportions of latitude to longitude (p. 80), we find the length and width as given by Marcian to be about 900 and 700 Roman miles respectively, his own figures being only 73% and 55% of those figures. On the other hand, although the distance for sailing round Britain cannot be checked from Ptolemy, it looks extremely long.

Marcian warns readers of the fallibility of sea measurements. He writes: 'I maintain that it is not easy to establish with accuracy the number of stades over every part of the sea. If the shore is straight and has no indentations or promontories, there is no problem in reckoning its measurement; but if it is full of bays and projections it is impossible to be accurate. One does not sail on a fixed route in the same way as one travels on a military road. Let us take as an example a bay which round the shore measures 100 stades. Anyone sailing close to the shore will find he covers fewer stades than the man walking along the beach, though there will not be very much difference; but the difference will increase as he traces the arc of a shorter circle.'[53]

BYZANTIUM AND LATER SEA CHARTS

Although we have no original Byzantine maps, there are Byzantine *periploi*, which followed in the tradition of earlier Greek ones as applied to the shores of the Byzantine Empire. In modern writings these have often been called 'portolans', but this is a name better reserved for the late medieval and Renaissance sea-charts, whose Italian name points to the importance in those periods of Italian cartographers. Geographical writers like Marcian may have drawn straight lines on their copies of maps to help them calculate distances; and these might correspond to the loxodromes which all except the earliest portolans have. But despite careful investigations, no direct line of descent has been able to be established between ancient *periploi* and these portolans.

It is of interest to see what features of the *periploi* could have been incorporated into portolans, so as to turn description into map form: (a) measurement: the *periploi* sometimes reckon in days' sail, sometimes in stades; as has been seen, the one could arbitrarily be converted to the other; (b) compass directions: only occasionally given, as 'after sailing for 6 stades you will see a promontory stretching out towards the west';[54] (c) information on world cartography: Marcian summarizes views from Eratosthenes onwards;[55] (d) information on harbours, moorings and beaches: some of this could have been incorporated into maps, with or without symbols; (e) trade etc.: such information often found its way into texts accompanying maps.

PERIEGESES

In addition to these prose descriptions, there were Greek verse *periegeses*, i.e. world guides. Dionysius Periegetes ('the guide') wrote between AD 117 and 138, in Alexandria, a *periegesis* of the known world in hexameters, based chiefly on Eratosthenes' map with little regard to subsequent cartography.[56] It was illustrated with a map by the time of Cassiodorus (p. 155). Dionysius describes the oikumene as sling-shaped, i.e. oval, and deals first with the Ocean, then with the continents, Asia being most fully described. He brings in mythology and refers to trade or migration; but on Britain, for example, he has next to nothing to say. The coasts are the predominant interests, but the Nile is described in outline as far as Syene (Aswan). The eulogy of the R. Rhebas in Bithynia has suggested that Dionysius must have come from there, since this was a river of no great importance.[57]

> Next live Bithynians in a fertile land,
> Where Rhebas summons forth his lovely stream,
> Rhebas, who travels by the Black Sea's mouth,
> Rhebas, the fairest water in the world.

The work became very popular and was translated into Latin verse by Avienius (p. 141) and Priscian. Another Dionysius of the second century AD, son of Calliphon, wrote a verse description of Greece in iambics, which includes a few measurements.[58]

Whereas Dionysius Periegetes is capable of sounding poetic, the same cannot be said of an iambic work of some 200–250 years earlier, the *Periegesis* of ps.-Scymnus.[59] The original Scymnus of Chios[60] wrote *c.* 185 BC a lost prose work of the same type. The verse *periegesis*, addressed to King Nicomedes III or IV of Bithynia, is fragmentary in its second half. Entries refer principally to coastal areas. The section on the Straits of Gibraltar may be rendered:

> The opening to the Atlantic Sea
> Is stated to be fifteen miles in width.
> The lands nearby are the extremities
> Of Libya (Africa) and Europe . . . .

It is possible that the preservation of such works is due to their having been used as school geographies.

Finally, an epigram has helped to date a *periplus* writer of whom we possess only a fragment. This is the *Periplus of the Euxine* by Menippus of Pergamon,[61] of which the first part is preserved; Marcian tells us there were *periploi* of the Inner Sea by Artemidorus of Ephesus (*fl.* 104 BC) in eleven books and by Menippus in three books, and that after the Black Sea Menippus gave a *periplus* of the Mediterranean. The dating of this writer is from an epigram by Crinagoras of Mitylene,[62] who visited Rome in 25 BC: 'I am preparing a sea voyage to Italy, to see friends I have not visited for a long time. I need a *periplus* to guide me to the Cyclades islands and ancient Scheria. Dear Menippus, you who know all geography, and have written a learned tour, please help me from your book.'

CHAPTER X

MAPS IN ART FORM

GREEK COSMOGRAPHY

The idea that maps or something approaching them were works of art and could best be portrayed in art form is very old and persisted throughout antiquity. In Homer (p. 55 above) it is the craftsman god Hephaestus who makes for Achilles a shield out of one gold plate, two tin and two bronze, one of the bronze plates being engraved with 'innumerable subjects, planned by a clever mind'. The heritage from Homer's artistry, related to a shield, is twofold. On the one hand the use of a shield map recurs under the Roman Empire with the Dura Europos shield (p. 120). This in itself contains not only roads and a coast in plan, but also an art element in the portrayal of ships, coloured and enlarged out of scale, to fill the empty spaces on the Black Sea. Just as in medieval and early Renaissance maps the *horror vacui* often led to artistic flights of fancy and irrelevant drawings, so in a land route map of this type the paintings on the sea had a similar function.

The other heritage lay in the realm of celestial cartography, which as it was well developed in Classical Greece (p. 21) may well have influenced the artistic representation of terrestrial mapping. The sky could be mapped on a flat surface or a globe. On a flat surface, it would appear that the artistic approach was quite as often used as more mathematical and scientific attempts at accuracy. The popularity of artistry is attested by the Latin translations, at various periods, of the extant verse rendering by the Alexandrian poet Aratus of Soli (*c.* 315–240 BC) of the *Phaenomena* of Eudoxus of Cnidos.[1] Although the description of constellations is poetic, one can clearly see that the animals or other figures which the constellations were thought to depict were themselves drawn in elaborate outline. It seems that Hipparchus of Nicaea (*c.* 190–120 BC) tried to combine the mathematical with the artistic approach, listing the exact latitude and longitude for eight hundred stars and criticizing Eudoxus if he had attributed an unsuitable part of a human or animal body to fit the name of a particular constellation.

This tradition appears in extant artistic form in the 'Lion of Commagene',[2] a bas-relief of 98 BC which depicts the royal horoscope of Antiochus, King of Commagene (north Syria). A lion, one of the signs of the zodiac, is shown with stars outlining parts of its body, and with three planets above it.

The artist who gave expression to Aratus' celestial sphere depicted the myth of the giant Atlas, supposed to hold up the earth. The Farnese Atlas,[3] in Naples, a Roman copy of a Hellenistic original, has a variant on this legend, in that the giant is holding up the sky. On the surface of the sphere are carved the figures representing constellations. The sphere is inclined on its axes, and has on it the equator, the tropics, the arctic and antarctic circles, and the two colures.[4] There are no constellations in the antarctic areas, invisible from the Mediterranean.

COINS AND MAZES

In Greece and Rome coins were often designed both as objects of art and as vehicles of propaganda; and stylized map-like features are sometimes portrayed. Harbours are shown on Greek coins from Messana (Messina) and on Roman coins. The early name for Messana was Zancle, 'sickle', referring to its sickle-shaped port. This is shown in ground-plan, with some protrusions thought to represent harbour buildings. A Roman coin of the Neronian period gives a ground-plan of Claudius' harbour at Portus, about 3 km north of ancient Ostia.[5] The two jetties are shown as arcs of a circle, and at the mouth of the harbour is the statue of an emperor, who must be Claudius or Nero. There is also a coin of Trajan which shows Trajan's hexagonal port in the same area, *c.* AD 113, with possibly identifiable buildings on each side.[6] With this type of miniature art it is often impossible to be sure of detailed and correct map interpretation.

An unusual coin type is to be found in an issue minted by Memnon of Rhodes, who during the expedition of Alexander the Great was appointed by Darius V temporary satrap of Ephesus.[7] The interesting feature of these coins is that they clearly portray, though only in rough form, some sort of relief map of the hinterland of Ephesus. Nothing similar has so far been found on

Roman coins. On a much larger scale of mapping, if they can be so classified, are the coins of Cnossos in Crete which portray a labyrinth. Mazes or labyrinths go back to remote antiquity, and in mythology the most famous was the one at Cnossos from which King Theseus, with the help of a ball of thread, is said to have rescued the young men and women of Athens. Early mazes appear both in square and in circular form, and these Classical coins of Cnossos show specimens of each.

The tradition of labyrinth plans lasted not only throughout antiquity (two early examples are from the 'palace of Nestor' at Pylos, of the Mycenaean period, and from the Athens acropolis, of the Classical period) but into the Middle Ages and beyond. In a number of cases these were associated with either Cnossos or Troy. In the first place, as at Cnossos, the maze concept may have originated in the labyrinthine passages of the early palace. But in the only extant Etruscan example, a vase found at Tragliatella with a circular maze,[8] there is a clear reference to the *lusus Troiae*,[9] which may or may not in origin have meant 'game of Troy'; it is described by Virgil as having labyrinthine movements.[10] This vase is inscribed TPVIA, Etruscan for Troy. It has been shown that the connection with Troy can be traced even to comparatively modern times.[11] An example of naive art form, a square maze drawing from the first century AD, comes, with numerous other graffiti, from Pompeii. Its connection with Cnossos is apparent in the inscription LABYRINTHVS: HIC HABITAT MINOTAVRVS.[12] The theme of Theseus killing the Minotaur, surrounded by a square maze, appears in mosaics of the late Empire. Two of these are strikingly similar: one in the Kunsthistorisches Museum, Vienna, and one in the Salzburg Museum, excavated from the Loig villa and considered to be early fourth century AD.[13] The tradition of maze representations goes on into the period of Christian mosaics.

SOUVENIRS

Bronze was a regular material for the engraving of maps, but it is also a material of considerable utility and beauty, and, as today, bronze vessels were sold as souvenirs. Some of these were sold to Roman soldiers manning Hadrian's Wall. The two chief extant specimens of these are vessels known as the Rudge Cup (already mentioned on p. 124 above),[14] and the Amiens patera.[15] They bear

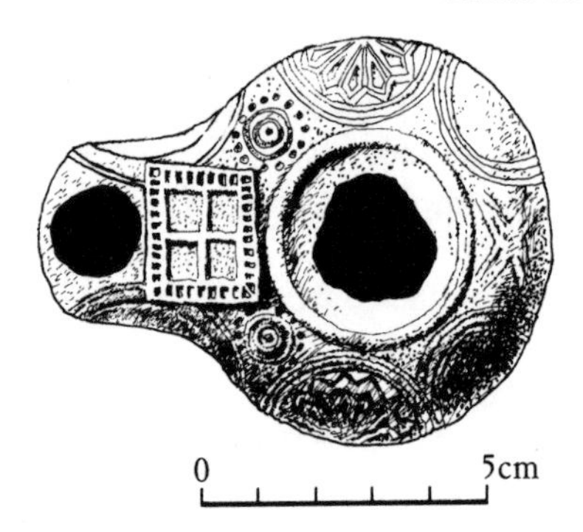

Fig. 27. Roman oil lamp from Samaria-Sebaste, early 4th century AD, with Roman fort in plan.

elaborate engravings which should not be classified as utilitarian, i.e. as a partial itinerary, contrary to what applies to the Dura Europos shield (p. 120). Rather they portray a part of Hadrian's Wall shown in stylized form with castellations.

The other medium which may have been used for military souvenirs was pottery. A Roman oil lamp from Samaria-Sebaste, Palestine, of *c.* AD 300–330, shows what is evidently a Roman fort in plan, with barrack rooms round the sides and with intersecting roads.[16] Two L-shaped road stations appear in plan on a fragmentary lamp from the Ophel. If such plans are not souvenirs, one could imagine them as linking the lamps to a particular site.

PAINTINGS AND MOSAICS

No painted wall-maps from the Graeco-Roman period are extant, but the earliest Roman map of which we hear seems to have been painted. In 174 BC Tiberius Sempronius Gracchus, father of the Gracchi, wished publicly to thank Jupiter for his victory over the Carthaginians in Sardinia. He therefore dedicated to him, in the temple of Mater Matuta[17] in Rome, a tablet consisting of a map of Sardinia together with an inscription and pictures of his victories. Such a map would probably not strive after accuracy, but would have a religious and propaganda impact. Two perspective views of map-like appearance are to be seen in manuscript miniatures of Virgil, *Aeneid*, on codex Vaticanus latinus 3225 (*c.* AD 420). On folio 31v is a ship entering a large harbour, and by its side the triangular island of Sicily, with eight buildings representing cities.

On the left side of folio 27r is a rough perspective of the same type which illustrates five Aegean islands.

Mosaics, on the other hand, which have already been mentioned in connection with mazes, were used extensively in later times for artistic maps and plans. Colour not only helped the artistic effect, but could, as on the Madeba map (p. 151), bring out a landscape feature. For cartographic purposes very small pieces of mosaic had to be used, and the technique of composition was obviously very specialized.

The use of this medium for maps occurs in its developed form, from what we know, only very late. There are, however, several earlier mosaics which contain a map-like element. A mosaic found at Mérida, Spain, in 1966, together with statues of AD 155, is thought to represent an allegorical view of the heavens in the upper portion and the earth in the central and lower portions, entirely in anthropomorphic form.[18] *Caelum*, 'heaven', is at the top, with *saeculum* on one side and *chaos* on the other. Below them are *polum* (= *polus*), points of the compass and winds, clouds, mountain, snow, thunder, nature, seasons, then geographical and other personifications: Nilus, Euphrates, *portus*, *pontus* (probably for Pontus Euxinus, the Black Sea), *oceanus*, *pharus* ('lighthouse'), *navigia*, together with *tranquillitas* and *copiae* (armed forces). It is to be noted that the Ocean is on the outside, near the lower left corner, but it is not possible to work out any overall topographical pattern.

The mosaic at the temple of Fortuna Primigenia, Praeneste (Palestrina), attempts to give a general pictorial impression of the Nile Valley.[19] The river appears as a body of water surrounding all the buildings shown. Probably of the second century AD, the mosaic contains Greek names of typical Egyptian animals. Obviously the buildings, shown in perspective, could be real ones; but it seems doubtful whether one can claim it as a semi-oblique type of map.

The ruins of Rome's port of Ostia have revealed, among many other mosaics, a number connected with transport, which was one of Ostia's chief concerns as an entrepôt. One of these has map-like features, viz. that in the forum of the Corporations (shortly after AD 196)[20] which housed shipping offices. The mosaic, in black, grey and white, shows a river spanned by a pontoon bridge resting on three vessels. This river has three branches which could be either tributaries or distributaries. On either side of the pontoon bridge is a gateway surmounted by military trophies in elevation. Three suggestions have been made for its location: the Nile, the Rhone and the lower Tiber. Of these, the Nile seems by far the most likely, because of the military trophies and the known corn trade from Egypt to Ostia and Rome. One can therefore think of the three distributaries as a simplification of the traditional seven, and locate the pontoon bridge between Memphis and Babylon (Old Cairo). The other type of transport depicted at Ostia which attempts to give a plan is the guild meeting-place of the *cisiarii*[21] (donkey-cart drivers), which can be seen to the left of the road leading from the modern entrance to the museum. A very stylized square castle, such as did not exist at Ostia, in part ground-plan, part elevation, is surrounded by *cisiarii* with comically named donkeys, such as Podagrosus ('gouty').

The mosaics from Roman north Africa which show buildings normally have them in elevation. But one temple scene from Carthage, known as the Offering of the Crane, tries not very successfully to combine ground-plan and elevation of a temple.[22] Another type of mosaic somewhat akin to the map is the pictorial representation of the zodiac. This occurs in mosaics from north Africa and Palestine, the former having successive signs of the zodiac clockwise and the latter anticlockwise. It is doubtful, however, if such representations can be thought to fall within the realm of cartography.

There are two colourful and attractive mosaic representations of the early Byzantine period, one of a Christian idea of the world, one of the Holy Lands. The former is at Nicopolis in Epirus, the new Roman 'victory city' founded by Augustus to celebrate his victory at the battle of Actium (31 BC).[23] It is *in situ*, and was set up by Archbishop Dometios, a Greek form of the Latin name Domitius, in the sixth century AD. Like the world maps of Ephorus and of Cosmas Indicopleustes, this world outline is rectangular. On the mosaic are Greek verses which may be rendered:

> Here you can see the boundless ocean run
> Carrying in its midst the earth, wherein
> All that can breathe and creep is here portrayed
> Using the skilful images of art.
> Noble archpriest Dometios founded this.

In actual fact there are no countries as such portrayed: rather the mosaic shows the traditional Ocean flowing round the earth, coupled with a recollection of Genesis on the creation of the earth.

Finally in this medium comes the most elaborate example of all, the Madeba mosaic map.[24] We can with certainty classify it in the artistic group, since mosaic lends itself to an attractive art form, and of this the Madeba mosaic is a prime example. The site is the Byzantine church at Madeba or Madaba, in Jordan, south of Amman. Its date lies between 542 and the death of Justinian in 565. As planned, it could have been seen by the congregation through a screen, with the lettering the right way up. It is well preserved but incomplete. It was discovered in 1884, having been covered over, and was treated and recorded in 1896.

The most surprising aspect, perhaps, is the enormous size. It is reckoned that in its original layout it could well have extended to a rectangle of as much as 24 × 6 m. Its object was to portray the Bible lands, so that it must have extended approximately from Egyptian Thebes to Damascus. Such a coverage would have required about two million *tesserae* (mosaic squares), and must have taken a very long time to prepare. The chief extant section now measures about 10 m 50 × 5 m. It is evident that for the two main colours, green–blue, used for mountains etc., and red, used for buildings representing towns, care was taken to provide a number of different shades; other colours too are quite well represented.

To most Christians of the period, Jerusalem was the centre of the Christian world in the geographical as well as the religious sense. True, scholars have been too keen to apply this rule universally, and to assume, for example, that the centre of the Ravenna Cosmographer's mental map was at Jerusalem, whereas to the present writer Rhodes fits better. But in the case of the Madeba map, despite incomplete survival, the treatment of Jerusalem proves that, though not central, it was the important place. It was purposely given an exaggerated size, with a scale approximately 1:1600 as compared to 1:15,000 for central Judaea. The inscription on it, which starts 'the holy city', shows that it was intended to convey Biblical reminiscences. It is shown as an oval walled city having its principal gate on the north. The main street is depicted with two colonnades, one the right way up, as far as all the lettering goes, and one upside down, as is the Church of the Holy Sepulchre.

We are fairly well informed on the topography of Byzantine Jerusalem, and numerous features can be identified, such as the New Church of the Theotokos (Virgin Mary), consecrated in 542, and the Damascus Gate with nearby column.

The coast of Palestine and Egypt, on the Madeba map, is shown with less change of direction than it actually has, so that the orientation of the map varies: in the Palestine section it has approximately ESE at the top, with the Mediterranean at the foot of the map, while in the Egyptian section it has approximately south-east. Thus the point of observation for each is the Mediterranean, as not infrequently in ancient map-making. The designers of the mosaic were obviously keen to include as many place-names as possible, mostly drawn from the *Onomasticon* of Eusebius (*c.* 340), so that in a number of areas it has a crowded appearance. Not only that, but Biblical explanations of several places are given, e.g. 'Ailamon, where the moon stood still one day in the time of Joshua son of Nun'; 'Rama: a voice was heard in Rama'; 'Zabulon shall dwell by the sea and its border shall be unto Sidon'.[25] The section which contained Egypt is not so well preserved, but we can see that despite the late date there are still villages called after the name of their Greek or Roman owner, such as 'the [village] of Nicias'.

Finally it should be noted that the conventional signs representing places or churches bear a strong resemblance to those in the Peutinger Table, suggesting some coherence which may have been established under the late Roman Empire. But the colouring in mountainous areas is more varied than that of the Peutinger Table, including red, pink, green and dark brown, possibly designed to depict vegetation and heights. A fragment evidently now lost may have pointed to a more sophisticated technique, a simplified form of contouring.

Whether the Madeba map was an isolated example or whether there were other Byzantine churches in Biblical lands with something similar destroyed by invaders, we do not know. It has been seen above (p. 128) that there was a tradition of itineraries for pilgrims to the Holy Land, and from areas like Byzantium such pilgrimages clearly continued as long as journeys were not too unsafe. On the one hand Justinian was keen to promote the incorporation of large and attractive mosaics in ecclesiastical and imperial buildings, as may be seen from Constantinople, Ravenna and elsewhere. On the other hand neither from extant remains nor from literary or inscriptional evidence can we infer that maps, whether in art form or not, were very common.

To some extent this map of Biblical lands foreshadows the world maps of the Middle Ages. Although these, such as the Hereford world map, attempted to cover the whole known world, they too had Jerusalem as their prominent feature. The tradition of brief explanations of Biblical topography also continued, as can be seen even from the utilitarian Peutinger Table (p. 113). But this map is thematic, whereas the Mappaemundi are general. They were circular or sometimes oval: the Madeba map, when complete, was neither of these, but roughly rectangular. It has no animals or representatives of mythical tribes as have these medieval maps. As has been seen, the orientation is only an approximation to the medieval concept of east, as the holy direction, at the top. Scale is obviously a problem in such a challenging material as mosaic, but some attention has been paid to it. No other extant Byzantine map is in any way comparable.

A LOST MEDIEVAL MAP

Finally in the Middle Ages we have a poetic record of a lost artistic world map. This was set up on the bedroom floor of Countess Adela at Blois, and is described in detail in the Latin hexameter verse of Abbot Baudri (*c.* 1100).[26] It should not be thought of as a poetic fantasy, but as an attempt in verse, in the style of ancient laudatory poetry, to describe a wealthy patron's work of art. Its features included mountains and rivers: thus in Italy Vesuvius, Monte Gargano and Monte Cassino were included. One detail shows advanced conservation technique: Baudri mentions that, to prevent the map deteriorating through dust, it was covered with glass. But the poet is not so explicit about its materials: it seems to have been either on silk covering a marble floor or incised in marble with painting added.

CHAPTER XI

THE DEVELOPMENT OF PTOLEMAIC MAPS

The legacy of Ptolemy's *Geography* lasted well over thirteen hundred years. As mentioned above (p. 80), he only once says that for his *Geographike Hyphegesis* (Manual of Geography) he has had maps made. Chapter headings in manuscripts such as 'Map 1' could perhaps be later interpolations. Elsewhere Ptolemy criticizes Marinus' map-making in detail. J. Fischer believed that the extant maps are derived by repeated re-copying from maps of the second century AD.[1] This theory was attacked by L. Bagrow[2] and others, but still has its devotees today. We do not hear from any ancient sources except Agathodaimon (p. 80) of a map relating to Ptolemy's *Geography*, though Cassiodorus may well be thinking of one (see below). Agathodaimon mentions only drawing a world map from it; we do not know whether he also drew regional maps. If neither he nor others did, readers could only visualize Ptolemy's topography if the work had maps or if they also possessed Marinus' work, unless the co-ordinates can be considered to have offered sufficient mental image. This could explain why Marinus' treatise long continued in circulation despite Ptolemy's criticisms: it is even mentioned in Arabic works.

LATER GREEK AND LATIN USE OF THE *Geography*

Pappus of Alexandria (*fl.* 379–395) is quoted by the Armenian geographer ps.-Moses of Chorene as having 'followed the individual map [or sphere] of Claudius Ptolemy'.[3] Fischer took this as proving that Ptolemy maps existed in the time of Pappus, whom he wrongly dated to one hundred years after Ptolemy. Ziegler, however, points out that the *Chorography of the oikumene*, as Pappus' title may be rendered, was probably a complete reworking of Ptolemy,[4] and Bagrow maintains that the Armenian word involved means 'sphere', not 'map', and that Pappus 'described the universe on the basis of the sphere made by Ptolemy'; though the title seems to rule out such a suggestion.

Some have thought that Marcian (p. 141) must have worked from regional maps based on Ptolemy, whether composed by a predecessor or by himself. It would certainly have helped him enormously to have had a world map at least, as has been seen in connection with north Britain. But the other source which he quotes apart from Ptolemy, the *Measurement of Stades* by one Protagoras, who presumably lived after Ptolemy, could have made maps less necessary for him. This mathematical geographer could have obtained figures of distance from Place A to Place B by reducing longitude by the cosine of the latitude.[5] Marcian's own contribution evidently lay in checking known distances against Protagoras' conversion of Ptolemaic co-ordinates.

Cassiodorus (*c.* 487–583), who was minister to Theodoric, King of the Goths, in Italy, advocated the study of Ptolemy. He founded a monastery at Vivarium near Squillace, south Italy, which clearly possessed a manuscript of the *Geography*, presumably in Greek, with or without maps. After commending first cosmography, then works on Constantinople and Jerusalem, he writes: 'Learn up the *pinax* [map] of Dionysius (Periegetes) . . . . Then, if you are inflamed by pursuit of this noble form of knowledge, you have a codex of Ptolemy, who so clearly drew [or described: the Latin is *tam evidenter descripsit*] all places that you may judge him to have been almost a native of every region. In this way, though established in one place as monks should be, you may mentally digest what one man's travel has collected with the greatest toil.'[6] In a letter from King Theodoric composed by Cassiodorus, the writer reminds Boethius that among the works which the latter had translated into Latin was the astronomy of Ptolemy.[7] Whereas Cassiodorus was a tolerably good Greek scholar, the Ravenna Cosmographer, although he quotes Ptolemy as an authority for the Roxolani and the R. Vistula, shows a lack of knowledge of Greek institutions by confusing him with Ptolemy I of Egypt or one of his successors.

ARAB USE OF PTOLEMY

Muslim geographers seem to have acquired and used copies of Marinus' map and Ptolemy's *Geography*, not necessarily complete and possibly in Syriac translation, as early as the eighth century. In the early ninth century al-Ma'mun, Caliph of Baghdad AD 813–833, set up an Academy of Science, which among other things produced a world map (lost) and 'improved tables', i.e. modernized co-ordinates. What is meant by the latter may clearly be seen from the *Kitab Surat al-Ard* of al-Khwarizmi (d. after 850).[8] This is thought to have been derived from a Syriac text and maps through the medium of an Arabic world map, the latter being either by al-Khwarizmi himself or by Ma'mun. Text and maps are preserved in a Strasbourg manuscript,[9] but the maps are thought to be later than al-Khwarizmi. They include one of the Nile valley, with south at the top in contrast to north on Byzantine Ptolemaic maps.[10] Two main sources of the Nile, from the south, and one subsidiary source, from south-east, are shown, and after many meanders it forms a delta with six distributaries, the westernmost near Alexandria. In the text the places are arranged according to their *klimata*,[11] starting with *Klima I* in the south, and are given longitudes and latitudes as in Ptolemy, in an Arabic form of the Milesian numeration.[12]

Some places are Ptolemaic, and here the co-ordinates more often disagree than agree, some non-Ptolemaic. It had been carefully revised so as to include new places. Kairouan, for example, did not exist in Classical times. We can see from the context that the latitude is not correct but is worked out in relation to adjacent Ptolemaic places, no doubt by reference to the Ma'mun map.

Ptolemy	*al-Khwarizmi*	*Longitude*		*Latitude*	
		Ptol.	*al-Kh.*	*Ptol.*	*al-Kh.*
Kulkul	Kulkul	28°30′	28°30′	31°15′	31°
—	al-Kairuwan	—	31°	—	31°40′
Bulla Regia	Rigiya-on-sea	30°40′	31°40′	31°30′	31°30′

The fact that some Egyptian places are entered doubly is thought to be due to revisions being incorporated without the original entries being deleted. There was in the time of al-Mas'udi (d. *c.* 956) at least one copy of Ptolemy's *Geography* with coloured maps containing according to him 4530 cities and over two hundred mountains. As the names of seas (and presumably other place-names) were in Greek, he says, he could not read them. But clearly we are dealing with regional maps and not merely a world map.[13] Al-Battani of Raqqa, Syria, included lists of towns with revised co-ordinates,[14] compiled about 901, in his astronomical tables.

Al-Biruni (973– after 1050) used the same method as is thought to have been used by the geographical mathematician Protagoras (p. 155) to calculate distances from Ptolemaic co-ordinates. In all, about 12,000 place-names are given co-ordinates of Ptolemaic type by various Arabic geographers, and these are in course of being collated at the American University of Beirut.

BYZANTIUM

In the late thirteenth century it was Maximus Planudes (*c.* 1260–1310) who in Constantinople revitalized Ptolemy's *Geography*.[15] He was a learned monk at Chora monastery, whose church with fine mosaics is still preserved as a museum on the outskirts of Istanbul. Unlike most Byzantine scholars of the period, he had a good knowledge of Latin. He was also interested in Greek astronomical and geographical works; he revised the *Phaenomena* of Aratus and collated old manuscripts of Strabo. He searched for manuscripts of Ptolemy's *Geography*, and his search was rewarded in 1295, but it was not as exciting as he had hoped. As he explains in a letter and in some verses,[16] after at last finding what he knew was a neglected work, he was disappointed to discover that it had no maps. He therefore set about providing maps to accompany the text. We almost certainly possess the manuscript which Planudes then encountered, since codex Vaticanus graecus 177, of the late thirteenth century, indicates that he was its owner. The only reference to a drawn map in that manuscript is the note inserted by Agathodaimon. But it seems not unlikely that Planudes had heard or read of maps earlier extant. For example, Vaticanus graecus 191 is a good manuscript of the same period or a little later. It has no maps, yet there is a note to the effect that there are twenty-seven maps instead of twenty-six, the tenth map of Europe having been divided into two. The obvious explanation seems to be that this is a note copied from a preceding manuscript which had maps.

The emperor (1282–1328) who acted as patron of map-making based on Ptolemy was Andronicus II Palaeologus. He was impressed by Planudes' maps and asked Athanasius, retired Patriarch of Alexandria, who from 1293 to 1308 was living in

Constantinople, to have made for him a special copy of the *Geography* with maps. There seems little doubt, despite the lack of any dedication, that this is the famous codex Vaticanus Urbinas graecus 82, reproduced in facsimile with extensive commentary by J. Fischer.[17] It came to the Vatican in 1657 from the library of Guidobaldo, Duke of Urbino, who is thought to have inherited it from his predecessor, Duke Federico da Montefeltro, and must be considered one of the world's masterpieces of cartography. It measures 57 × 41.5 cm, and is richly illustrated. The colours on the maps are the traditional ones of blue for sea and rivers, outlines of coasts and rivers being drawn rather schematically, brown for mountains, usually shown as straight ridges or gently curving ridges; these colours are very well preserved. The world map, at the end of Book VII, is on Ptolemy's first projection (simple conic), with the oikumene surrounded by a conic frame. Outside this frame are winds, shown as faces blowing horns, and signs of the zodiac in red circles. There are 26 regional maps, comprising 10 of Europe, 4 of Libya (= north Africa) and 12 of Asia; these are interspersed in the text of VIII.3-28. As intended by Ptolemy, they are rectangular, each being given the proportion of latitude to longitude appropriate to the region. Meridians appear every 5 degrees, half-degree marks being shown in the outer frame. Parallels appear at every quarter-hour of maximum sunlight, a method adopted by Ptolemy in Book VIII; whereas in the marginal frames degrees of latitude are given, as in his co-ordinates. Place-names entered, whether civilian or military, have their centre indicated by a rectangle, which in the case of those considered important[18] is elaborated into one with castellated top. In a heavily populated area the large number of rectangles and names results in a somewhat cluttered effect. Special features like altars and columns are, as in the Peutinger Table, given conventional signs.

One set of conventional signs deserves particular notice. In areas of western Europe where the Romans had institutionalized native tribes of a province into *civitates*, Ptolemy's text, although it gives no indication of boundaries, allocates to individual tribes the places for which co-ordinates are given. Thus an extract from the British section reads as follows: 'Further south than these [the Damnonii] are the Otadini [= Votadini], in whose area are the following places:

Curia	20°10′	59°
Alauna	23°	58°40′
Bremenium	21°	58°45′ ’.[19]

Plotting such places on the map, given their longitude east of the Canaries and their latitude, presented no problem to the Byzantine scholars, despite their unfamiliarity with Britain. But in the absence of boundaries a method was needed for indicating the attribution of *poleis* (places, literally ‘cities’) to tribes. This was effected in the Urbinas by a system of small conventional signs, originating in astronomical writings, attached to each place-name in these provinces. We may therefore think of the Urbinas in these regions as the father of political maps, since the first printed editions of Ptolemy devised boundaries to replace the symbols. The existence of these signs does not provide evidence of copying from previous maps whose archetype was in or shortly after Ptolemy's time, since with a reliable text the whole visual system could have been reconstructed, probably with no maps, certainly with only a world map to act as basis.

Such a degree of accuracy was not followed in other Ptolemaic maps, apart from the Latin translation (p. 161) which was modelled very closely on the Urbinas. The two codices with maps whose text and format most closely resemble the Urbinas[20] are in Istanbul (Seragliensis 57) and Copenhagen (Fabricianus graecus 23, only a double folio preserved). These may have been private copies made for Planudes and Athanasius. It is a pity that Planudes was not

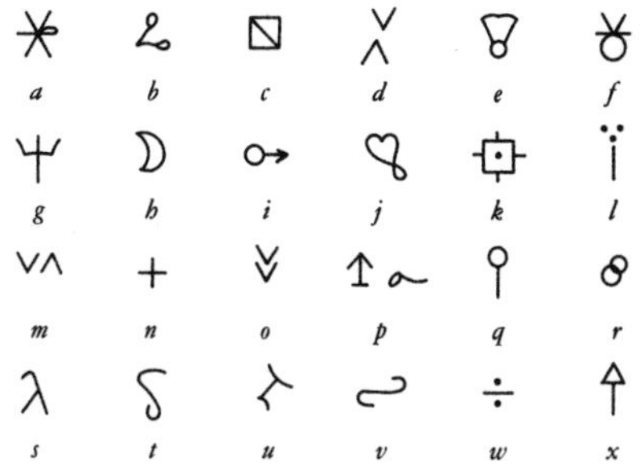

Fig. 28. Ptolemy: conventional signs for British tribes in Codex Urbinas graecus 82, Vatican: (a) Atrebates, (b) Belgae, (c) Brigantes, (d) Cantium, (e) Catuvellauni, (f) Coritani, (g) Cornovii, (h) Damn(on)ii, (i) Demetae, (j) Dobunni, (k) Dumnonii, (l) Durotriges, (m) Iceni, (n) Novantae, (o) Ordovices, (p) Parisi, (q) Regni, (r) Selgovae, (s) Silures, (t) Taexali, (u) Trinovantes, (v) Vacomagi, (w) Vennicones, (x) Votadini.

able to follow up his rehabilitation of the *Geographia* by a visit to the chief centres of learning in the West. Such a journey might well have resulted in the dissemination of manuscripts of a Latin translation a hundred years before it actually happened. As things were, it is premature to speak of a general renaissance of Classical cartography at this time.

Mention should also be made of Vatopedi 655, of the early fourteenth century. Vatopedi is an old and famous monastery on Mount Athos. The text of this manuscript, which includes Strabo and minor Greek geographers, is of some interest. But the maps have been shown to be poor copies of the Urbinas, obviously made before the latter found its way to Italy. Moreover the manuscript has been dismembered to some extent. The map of the oikumene, in simple conic projection, was abstracted by the Greek collector Simonides in the nineteenth century and is now in the British Library;[21] while several folios were said by Bagrow to be in a Leningrad museum.[22]

In addition to manuscripts with twenty-six or twenty-seven maps, which are known as those of Recension A, there are somewhat later manuscripts, known as Recension B, which have sixty-five maps. The earliest of these extant is in Florence, Laurentianus XXVIII.49, of the early fourteenth century. It is of smaller format and its whole arrangement of maps is different. Maps of Europe, in Books II–III, number twenty-five; of Libya, in Book IV, number eight; of Asia, in Books V–VII, number thirty-one. This gives a total of sixty-four, followed by a map of the oikumene in a simple conic projection. Conventional signs for towns tend to be different from those of Recension A, including from one to three towers. There are a number of other manuscripts with Recension B maps, dating from the fourteenth and fifteenth centuries, including some with 'modern' maps. One cannot say that the great increase in number of maps in B as compared with A improved the cartographic detail: on the contrary, the Urbinas maps of Recension A remain the most detailed and accurate.

LATIN TRANSLATIONS

Contact between Byzantine and Western scholarship was only sporadic until late in the fourteenth century. In 1395 Manuel II Palaeologus sent the scholar Manuel Chrysoloras (*c.* 1350–1415) to

Venice, whose help he was soliciting as an ally against Turkish inroads.[23] In Italy he made contacts with Classical scholars, and returned to Constantinople with his gifted student, the Tuscan Jacopo d'Angelo da Scarperia. Two years later they were back in Florence, where Chrysoloras had been invited for two years to teach Greek. Although they had brought manuscripts with them, there was an obvious need for the translation into Latin of such works as were unfamiliar in the West.

Chrysoloras, who translated Homer and Plato, himself intended to translate Ptolemy's *Geography*, but delegated the task to Jacopo d'Angelo,[24] who from 1401 was secretary to the Papal Curia. The Greek manuscript used by Angelo for his translation has not been identified. The translation was completed in 1406 and originally dedicated to Pope Gregory XII; this may be seen from the earliest manuscript,[25] whereas others have a dedication to Alexander V (1409–10). In his introduction Angelo praised the author of the *Geography* for drawing the earth on a flat surface while keeping the size of regions in relation to each other and to the whole earth relatively proportionate, and for incorporating a practical system of latitude and longitude co-ordinates. He was, however, no mathematician, and his comprehension of the mathematical content of the work was criticized later in the fifteenth century by Toscanelli and Regiomontanus.[26] For Ptolemy's title *Geographike hyphegesis* he substituted *Cosmographia*. This was an unfortunate choice, since it would most naturally be taken as describing the mapping of the universe. Nevertheless, it continued to be used in successive editions of his translation, and it was not until the late fifteenth century that the alternative title *Geographia* was substituted.

Angelo had not latinized the Greek wording on the maps, and may in fact have issued his translation without maps. Latinization was first executed in 1415 by Francesco Lapaccino and Domenico Boninsegni. They exactly copied the maps of Urbinas graecus 82, and their Latin maps are preserved in the extant Vaticanus latinus 5698. It has been conjectured that this manuscript, which contains only maps, was intended as a supplement to Parisinus latinus 17542, which has text only; but this cannot be proved.

The chief promoters of Ptolemaic cartography were Florentine humanists such as Palla Strozzi under Medici patronage. In 1434 Strozzi moved to Padua and continued his support from there.

Other areas of Italy too had their patrons: in Ferrara, for example, there was the Duke Borso d'Este. Church patronage was also important, particularly in the revision of Ptolemaic maps. Thus in the papacy of Pius II (Aeneas Silvius Piccolomini), 1458–64, cardinals saw to the supplementation of Ptolemaic maps by *tabulae modernae*, newly drawn maps to add to cartographic knowledge in selected areas of the world. Some of these, such as Scandinavia and Ethiopia, were beyond the reach of detailed Ptolemaic mapping, while others were within it but needed some updating.

One of the most famous names in the development of Ptolemaic cartography was Donnus Nicolaus Germanus (*c.* 1420–*c.* 1490),[27] a Benedictine monk of German origin at the Badia di Fiesole near Florence, who worked among others for the Vatican and for Borso d'Este. His greatest innovation was the use of trapezoidal projection for Ptolemaic maps. Ptolemy had said that he included his first as well as his second projection for the sake of those who through laziness were attracted to that easier method.[28] Donnus Nicolaus therefore began to think how he too could originate a new projection.

During the fifteenth and sixteenth centuries there was great demand for elaborately decorated Ptolemaic maps, especially in court circles. One of the most ornate, with magnificent calligraphy and illumination, is a copy made for Henri II of France and now in the Bibliothèque Nationale, Paris.

PRINTED EDITIONS

From the 1470s dawns a very different period, that of printed editions of the *Geography*. The earliest with maps were produced in Bologna (1477), Rome (1477–8) and Ulm (1482). The first printed text, that of Vicenza (1475), had no maps. Now vastly more copies could be produced and disseminated. The more they were used, the more they could be revised, criticized and ultimately replaced.

The Bologna edition[29] has the first maps to be printed from copper plates, a significant advance in the history of cartography. Unfortunately they are not as accurate or attractive as those of the Rome edition, the reason evidently being that the editors were rushing them out so as to anticipate their Rome rivals. The world map is on Ptolemy's first projection. Instead of the customary twenty-six maps of Recension A, there are twenty-five, parts of

Asia Minor being inserted in the tenth map of Europe. The Latin text and maps have many misprints, e.g. *n* for *m* in CLAVDII PTOLEMAEI | COSNOGRAPHIAE LI | BER PRINVS; and on the map of Britain ALBIVNINS | VLABERT | ANIA stands for *Albion insula Britannia*. Forests, as in some manuscripts, are represented as clumps of half a dozen or more trees. Town names are in capitals with a dotted circle to show the exact location; but on many maps, including that of Britain, only a selection is given. Sarmatia has RIPHEV. M. (*Riphaeus mons*) surmounted by Alexander's columns. Mythical attributions to Alexander's visits exist in Ptolemy's text, but there is here a confusion between his columns and his altar. From there south-west to the Carpathians are first one ridge of mountains, then in places up to four. The maximum length of daylight is given as well as latitudes.

Following hard on its heels was the Rome edition of 1477–8,[30] whose preparation clearly started long before that of Bologna. The promoter of this edition was Conrad Sweynheym, who about 1464 had set up at Subiaco the first printing press in Italy. Two letters from Domizio Calderini to Pope Sixtus IV give details of this. Referring to Sweynheym's activities from 1473 or 1474, he wrote: 'Calling in mathematicians [presumably to explain the projections], he taught how maps could be printed from copper plates. After spending three years on this labour he died.' The work was taken up by his associate Buckinck; but there was inevitable delay. The world map is on Ptolemy's first projection, and there are twenty-six regional maps. The lettering is in two sizes of very clear capitals, and the whole of the mapwork is carefully done. Altars surmounted by flames are large and prominent, and mountains have a distinctive inverse UUU shape. The area enclosing the Indian Ocean to the south is marked *terra incognita*.

The Ulm edition,[31] the first printed edition of the *Geography* to appear outside Italy, had maps designed by Donnus Nicolaus Germanus (p. 162), as we can read in the colophon. The world map is on Ptolemy's second projection, with a few additions north of the 63° latitude; these were amplified by him in subsequent recensions. Many of the maps are noticeable for the reversed И, possibly serving as a sort of trademark. The regional maps are in Donnus Nicolaus' trapezoidal projection, except that the map of Spain is rectangular and that of Egypt-Ethiopia is in the form of a double trapezium widest at the equator. In addition to forests, there are conventional signs for altars, shrines etc. Undoubtedly in certain respects the maps served an antiquarian purpose, being over a thousand years out-of-date for some considerations. Thus, for some regions, including Britain, tribal boundaries are shown, presumably inserted somewhat arbitrarily from a map which had conventional signs indicating distribution of towns to tribes. Britain also has far more entries than in the Bologna edition, with little inaccuracy. Whereas the Bologna edition labels its mountains, the Ulm edition puts wording, such as *taurꝰ mons* for the Taurus range, by the side of a range in minuscules. The result is that empty white spaces are all that is left on the ranges, which make them look rather like lakes. Those early editions were all printed in black and white, but could be hand-coloured.[32] Such hand-colouring could be done with a manuscript as model. In order, however, to provide for uncoloured copies, better solutions for showing mountains proved to be shading or hachuring.

PTOLEMY IN THE AGE OF DISCOVERY

The Portuguese discoveries sponsored by Prince Henry the Navigator had resulted by 1445 in the rounding of Cape Blanco and Cape Verde, and geographers began to take a new look at Ptolemy to see if Africa could be rounded to reach the Indian Ocean. Even if reports of early circumnavigations of Africa were unknown or disbelieved, the theoretical idea of an encircling ocean was extremely ancient. But the enclosing of the Indian Ocean by a strip of land extending from east Africa to the Malay Peninsula is found in Ptolemy's text[33] as well as in the late Byzantine and Renaissance regional maps and some of the world maps (p. 178). Eventually it was the voyages of Bartolomeu Diaz, more correctly Dias (1487), and Vasco da Gama (1497) that necessitated changes from Ptolemaic cartography. As in the north, these were sometimes effected by means of *tabulae modernae*, when the usual practice was to include the old map as it was and add a new one (though we still, for example, find the Ptolemaic shape of north Britain in a *tabula moderna* of Scandinavia, in the Rome Ptolemy of 1507[34]). In the intervening period there was some attempt at reconciling Ptolemy maps with the partial circumnavigation of Africa, as can be seen in the Martin Behaim globe of 1492.[35] But as marine discovery in the south and south-east progressed, maps broke away from Ptolemaic shapes, and Taprobane, thought not to be Sri Lanka because it was too big, was reallocated to an island off the Malay Peninsula.

As cartographers came to realize, Ptolemy's view of the Old World had its defects. The principal defect was that he had wrongly calculated the length of the Mediterranean and the east–west length of the oikumene in terms of degrees of longitude. His length from the Straits of Gibraltar to the Gulf of Iskenderun amounts to 62°, whereas the actual length is 41°40′. His length from the Canaries to Cattigara amounts to 177°, and to Thinae 180°, whereas from the Canaries to Hanoi is actually 124°. The dimensions of the Mediterranean were much better calculated on some fourteenth-century portolans, but many maps continued to use the earlier measurements even as late as the seventeenth century. Secondly, there are numerous misconceptions, both within and outside the limits of the Roman Empire, such as have been mentioned in Chapter V.

The legacy of the *Geography* played a vital part in the discovery, or re-discovery, of America, and editions of Ptolemy were amended to include the newly known lands.[36] Christopher Columbus and his brother Bartholomew, a chart-maker, had access to Ptolemaic maps. By extending the east–west length of Eurasia to 225°, which had been Marinus' figure, and then adding 28° for the lands discovered by Marco Polo, plus 30° to the east coast of Japan, Columbus concluded that the voyage to Japan from the Canaries, for which he wrongly added another 9°, would occupy only 68° longitude.[37] An edition labelled as Ptolemy's *Geography*, that of Waldseemüller (1507), actually has the first naming of America as a continent. It is to be found both on the map and in a text of the same year intended to accompany the map but probably written by Waldseemüller's associate, the poet and humanist Matthias Ringmann.[38] In this text Ringmann, if it was he, wrote: 'Since a further fourth part of the world has been discovered by Americus Vesputius [Amerigo Vespucci] . . ., I do not see why anyone should object to its being called, from its shrewd discoverer Americus, "Amerige", as if "Land of Americus", or "America", since Europe and Asia have derived their names from women.' It is now obvious that this cannot be 'Ptolemy's *Geography*', as Ptolemy had no conception of such a world. It was not to be long before the new discoveries and a general desire to break away from tradition caused a split between historical and contemporary cartography and an end to the practice of fathering world maps on Ptolemy.

So Ptolemy, having in Classical times influenced cartography and the recording of latitude, for example in portable sundials, had by the Renaissance come to be so esteemed that he had almost dominated mapping of the world, determined its exploration and therefore influenced its settlement. The advantages of his method were that one had detailed text and could have accompanying maps, plotting places by exact textual co-ordinates; that the regional maps were easy to construct, being orthogonal, and designed to reduce distortion to the minimum; and that the affiliation of towns to political areas could be indicated by conventional signs. The disadvantages were that his first world projection was an awkward contrivance even in Ptolemy's time and became more problematic with the southward extension of the known world, while the second projection had meridians whose curves were difficult to draw accurately; that rivers tended at best to be sketchily represented; that mountain ranges were inserted somewhat arbitrarily in the absence of detailed co-ordinates in the text, while individual peaks hardly appeared at all; and that coastlines were devised to link plotted coastal places rather than accurately surveyed. Moreover throughout the centuries there was a tendency to accord the *Geography* the type of infallibility accorded, for example, to Aristotle, so that it was a long time before editors had the courage to correct even features known to be incorrect. This attitude was, however, not that of the early Arab cartographers like al-Khwarizmi, whose aim was, by insertion of new material, to create a really up-to-date map. But these were often misunderstood by Europeans, and total map knowledge did not advance as a simple addition of the two. In fact they helped to compound Columbus' errors.

CHAPTER XII

FROM ANTIQUITY TO THE RENAISSANCE

THE LATE EMPIRE

There seems to be little doubt that with few exceptions the standard of cartography declined under the late Roman Empire. Assessment, however, is difficult, since some items known to have existed are lost, for example the geography or map which Alypius sent to Julian (Emperor 360–363) from Britain, while others are preserved only in late manuscripts. One example in the latter category is the Notitia Dignitatum,[1] a list, as its name implies, of civilian and military office-holders and administration, accompanied by maps of appropriate locations. Its dates are between 395 and 413, perhaps with later revision, and it is divided between Eastern and Western Empire. Even if the extant Notitia is not an official document but an unofficial copy, as some believe, one might expect it to have respectable maps, especially as there was a branch of the civil service in Rome at that time which dealt with maps. It was under the *comes formarum*,[2] though we know nothing of its activities. But instead of being influenced by such a department, the maps seem to have been compiled line by line from the text, by someone who had little topographical knowledge and who inserted place or province names in textual order. Perhaps this reflects greater difficulties of travel in the disturbed conditions of the late Empire.[3] It is even possible that they were invented in the Dark Ages: they go hand in hand with the other illustrations, which include very many emblems of the units described.[4]

An example of the method of compilation of maps is the British section. Inside a roughly shaped coastal outline are five provinces,[5] arranged one in the north, two in the centre and two in the south. The more easterly in the south, according to this map, is the province of Britannia Prima, which looks as if it were centred on London. But in point of fact we know that Corinium (Cirencester) was the capital of Britannia Prima, while London was probably the capital of Maxima Caesariensis. Somewhat similarly,[6] in Mesopotamia both Amida and Constantina are twice included, because each is twice mentioned in the list of military units. The same is true of Coptus in Egypt, and Upper Egypt follows neither topographical nor textual order. Sometimes, as in the section on the *comes Isauriae*, many forts are shown but no place-names. The section on the *comes Britanniarum*[7] has a single unnamed town on a rough map labelled BRITANNIA.

Obviously such defects could conceivably be due to repeated copying. But we can trace the history of Notitia Dignitatum manuscripts back at least to the tenth century.[8] There was at that time an illuminated manuscript of the work at Speyer Cathedral, on the Rhine above Ludwigshafen. It disappeared in the sixteenth century, but not before it had been copied several times. By comparing the two Munich versions (1542–50)[9] with other manuscripts at Frankfurt, Paris, Oxford and Cambridge (all fifteenth-century), we can see that they reflect the pictures of the tenth century fairly faithfully. The style of architecture is consistent with the maps having been contemporary with the text, or perhaps its revision. So it may well be that the maps existed from the start and were never intended as good topographical guides, but as outlines drawn by bureaucrats for other bureaucrats. They do contain some correct features, such as the course of the Euphrates and the position of Carrhae, notorious to the Romans as the scene of Crassus' defeat, east of its upper course. They do show

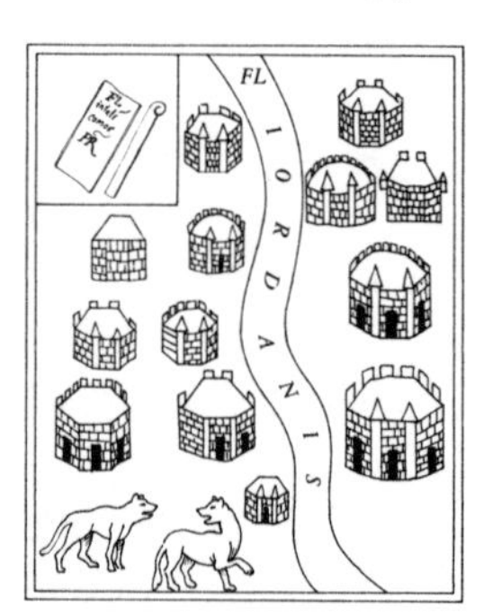

Fig. 29. Notitia Dignitatum: R. Jordan and towns under the Dux Palaestinae. The map is not realistic but compiled, like others in the Notitia, from lists: thus Aelia (Jerusalem), last line on right, is not near the Jordan and not orientated correctly in relation to other places.

that the outline of Britain was not always turned eastwards in its northern parts as in Ptolemy. But for factual information, which is often of great assistance to topographical research, we need in this case to turn to the text of the Notitia and not to its maps. In Britain, for example, the names and dispositions of all army units in the northern frontier area are there recorded; the information on the Saxon Shore forts has enabled the locations of these, with only one uncertainty, to be established;[10] and we learn of the existence of a government textile factory at a town only called Venta,[11] not entered on the maps, which must be Winchester (Venta Belgarum) or Caistor St Edmund (Venta Icenorum), near Norwich, rather than Caerwent (Venta Silurum). Likewise information on other areas of the Eastern and Western Empire can, if treated with caution, be of great assistance to historical and topographical research.

The two painted itineraries of which we know, the Peutinger Table and the Dura Europos shield, are products of the late Empire, even if the former had first-century AD antecedents. These and simple itineraries, such as the Antonine Itinerary and those prepared for pilgrims, are discussed in Chapter VIII. They were practical guides, and despite their shortcomings may have been almost as correct as preceding maps or itineraries, from which most of them were obviously derived. The same applies to *periploi* (Chapter IX), which evidently continued to be in list form.

From the first half of the fifth century AD we learn of a world map carried out under the orders of Theodosius II, Emperor of the East from 408 to 450.[12] The actual date is likely to be 435, since the phrase 'fifteenth *fasces*' in the cartographers' poem, which alone survives, should refer to the Emperor's fifteenth consulship. This poem is in Latin hexameters, and may be rendered:

> This famous work – including all the world,
> Seas, mountains, rivers, harbours, straits and towns,
> Uncharted areas – so that all might know,
> Our famous, noble, pious Theodosius
> Most venerably ordered when the year
> Was opened by his fifteenth consulship.
> We servants of the emperor (as one wrote,
> The other painted), following the work
> Of ancient mappers, in not many months
> Revised and bettered theirs, within short space
> Embracing all the world. Your wisdom, sire,
> It was that taught us to achieve this task.

The use of Latin may be explained by the fact that it was still at that time the official language of the Greek-speaking Eastern Empire. The wording relating to the two compilers is interesting as showing the division of labour. Whether they actually went round both parts of the divided Empire mapping it is doubtful. Their internationally vague language could merely mean that they updated a previous official world map. Since we have definite knowledge of only two such previous Roman maps which can be considered official, those of Julius Caesar and of Agrippa, it is likely, if so, to be the latter that they updated. As it was by then some 450 years since the compilation of Agrippa's map, revision was obviously much needed. Many new towns had been founded; some had disappeared, though if Theodosius' map was in this respect comparable to the Peutinger Table, it may have left them in. But in addition there had been enormous tribal migrations, so that many tribal names would have had to be moved. This was the last official map available in the West, and Dark Age geographers like the Irishman Dicuil (*fl.*825) used it as a principal source.[13]

A curious combination of windrose and world map is provided by a miniature in manuscripts of Ptolemy's *Handy Tables* (πρόχειροι κανόνες), an astronomical work.[14] It is circular with north at the top and with a division of the circle into twelve, with ten winds and North and South Poles. Thus far there is a close correspondence with Aristotle's *Meteorologica* (p. 28). But the central point does not, as in Aristotle and the Pesaro anemoscope, represent where the observer is, and there are selected topographical names added, such as Persia, Persian Gulf and several oceans. One unusual feature is a narrow rectangle running NE–SW and detailing places in the Nile Valley, as far as 'Lake of Meroe'.[15] At the bottom of the map is the entrance to the mythological underworld, with 'R. Lethe', 'R. Pyriphlegethon', 'Lake of Acheron'. The map-maker is likely to have been a Greek living in Upper Egypt in the fifth or sixth century AD. Despite being attached to an astronomical work there is nothing astronomical about the map.

BYZANTIUM[16]

Our information about Byzantine cartography is incomplete. Extant items belong to the earlier period, but the later period was conspicuous for the rediscovery of Ptolemy's *Geography*. The most detailed work is that of the Madeba mosaic map (p. 151). When this is studied in relation to Classical maps, the obvious difference lies in the greater size and detail of Jerusalem, and this emphasis was to have a pronounced impact on the world maps of the Middle Ages. The Holy City is not, however, central to the Madeba map as it is to many of these. One may imagine that in a less difficult medium than mosaic such maps achieved a greater degree of accuracy.

Christian world mapping was not fully developed, as far as we know from what remains, in the early Byzantine period. The Nicopolis mosaic (p. 150) is so undefined as to be only on the borderline of terrestrial mapping. Far more specific, though curious to our eyes, are the maps preserved in manuscripts of Cosmas Indicopleustes.[17] Cosmas (*fl.* AD 540) was an Alexandrian merchant who had had little formal education. His name Indicopleustes means 'who sailed to India'. Whether he actually reached India we do not know, but he did travel widely, and came to be converted to Christianity by a teacher in Persia. His extant work, in twelve books, is the *Christian Topography*:[18] his lost works are one on geography and one on astronomy.

Cosmas turned his back on what had since Eratosthenes been the accepted view that the earth was spherical.[19] True, there had been different views both as to the extent of the oikumene and as to the best way in which to represent it on a flat surface. Possibly Macrobius' revival of ideas propounded by Crates of Mallos had opened up a phase of criticism. Cosmas was a keen reader of the Bible and something of a fundamentalist. He held that the earth was flat and the universe box-like. Whereas John Philoponus, a contemporary of his in Alexandria and a commentator on Aristotle, took the regular view of a spherical earth, Cosmas denounces such views as coming from 'outsiders', i.e. pagans. Just as he rejects a spherical earth, so he rejects antipodes, using a sort of cartoon, with four giants standing on a small globe, to make fun of the idea.

The universe is shown, in the manuscript miniatures of Cosmas,[20] by a crude attempt to portray a three-dimensional object, as a vaulted box with convex top, something like the Ark of the Covenant (Exodus 21.20), which was ordered to be made of acacia wood with the dimensions $2\frac{1}{2} \times 1\frac{1}{2} \times 1\frac{1}{2}$ cubits, standing on four ringed feet. The upper section of Cosmas' universe represented God's realm, the lower section our world. In another diagram[21] the latter is portrayed as a flat rectangle with the Ocean flowing round it, somewhat as in the Nicopolis mosaic. Biblical ideas of the shape of the earth could easily be held to fit in with such a scheme. The oikumene is shown as a land-mass with four gulfs draining into the Ocean, north being at the top. In the north is the Caspian Gulf, which like a number of earlier geographers he did not think of as an inland sea. In the south are the Arabian Gulf (Red Sea) and the Persian Gulf, and in the west the Romaic Gulf (Mediterranean). Beyond the eastern border of the world is Paradise, from which flow four rivers, the Geon (Nile), the Tigris, the Euphrates and one called the Pheison, which although it is shown as also flowing into the Persian Gulf may in fact correspond to the Indus. Beyond the ocean, in addition to Paradise, is the 'earth beyond the ocean', inhabited before the Flood. As has been noted, the whole scheme is a curious mixture of Hellenistic and Biblical concepts.[22] On the Hellenistic side we have the great gulfs, the length of the oikumene, and the rectangular frame. Cosmas in fact does attribute a map of his type to Ephorus (*c.* 405–330 BC), whose universal history in thirty books constituted the chief source of Diodorus Siculus' history. Thus he was adapting a map of nearly nine hundred years earlier to incorporate Christian concepts, and it is interesting to see that Ephorus' work was still in circulation or at any rate available in the Alexandrian library. This particular map appeared in Book IV of Ephorus' lost history, where he was discussing what predecessors had said about Ethiopia.[23] He set up the fourfold scheme tabulated on p. 27 above, consisting in effect of two major areas, those of the Ethiopians and the Scythians, and two minor areas, those of the Indians and the Celts. He makes no attempt to provide for change due to tribal migrations.

The approach of Cosmas Indicopleustes is thus totally backward-looking, on the one hand to the Old Testament, on the other hand to Classical Greece. Although his method was unscientific, one may think of him as the forerunner of later cartographers who used maps as a vehicle for expounding their religious views. His approach may have influenced European mapping, e.g. that of Beatus (below), but does not seem to have influenced Arabic mapping. The respective contributions of each of these are briefly discussed in the following paragraphs.

THE DARK AGES

In the West too there were many old maps in circulation, but from the seventh century onwards we find a growth of the very simple outline world map known as T-O,[24] the O signifying the surrounding Ocean and the T the division between the three continents of the ancient world. It has been suggested that the T symbolizes the tau cross, a symbol both in the Old Testament and in early Christianity.[25] The earliest extant example of a T-O map is thought to be in a manuscript of Isidore of Seville at St Gallen, dated on palaeographical grounds to *c.* 700. Others occur in the works of Sallust, Lucan, Macrobius, Orosius and the Venerable Bede. Isidore is often quoted as the originator, but we may wonder if much earlier commentators did not invent the form. A number of those in manuscripts of whose works T-O maps appear have some connection with Africa, though whether this points to Classical or medieval origins is hard to say: Sallust wrote on the Jugurthine War in Numidia, Lucan on the civil war, including its phase in north Africa, whose geography he outlines. Macrobius (p. 174) was probably born in north Africa and had his own idea of the mapping of Africa; and Orosius fled from Spain to Africa in 414, incorporating in his *Historiae adversum paganos* a synopsis of world geography.[26] The earliest type, however, may have been a simple T-O without geographical details, and the contribution of Isidore in his *Etymologiae* was to impart a Biblical flavour to the bipartite division. Asia, Africa and Europe were regarded as the inheritance of Noah's sons Shem, Ham and Japhet. Asia, he continues, 'is bounded on the East by the sunrise, on the South by the Ocean, on the West by the Mediterranean, on the North by Lake Maeotis [the Sea of Azov] and the R. Tanais [Don]. It contains many provinces and districts whose names and topography I will briefly describe, beginning with Paradise . . . Paradise is a place situated in the eastern parts . . .'. Hence we find *Meotides paludes* (Maeotis marshes) and Paradise in a rectangle with four rivers. This develops with Beatus of Liebana (*fl.* 776–786)[27] to a rectangular type oriented to the east, with not only Paradise but a fourth continent, which Beatus considered as inhabited and as one of the corners of the earth visited by the apostles. Legends on these maps associate the fourth continent with the Antipodes, though there is no connection with Crates' orb. On the standard tripartite version there are variants, including (a) one in which the continents are labelled the wrong way round, whether intentionally or not, (b) one incorporating measurements of continents; sometimes Pliny's figures are mistakenly multiplied instead of being treated as thousands (with line above to multiply by 1000) plus units.

The other type of small world map found from the eighth century or before is the circular one with six zones roughly representing climatic belts. The antecedent of this was probably the world map of Macrobius, perhaps to be equated with the Macrobius who was proconsul of the province of Africa in 410. In his *Commentary on Scipio's Dream*, a Neoplatonist exposition of the *somnium Scipionis* contained in Cicero's *De re publica*, Macrobius follows Crates (p. 36) and Cicero in postulating a belt of water in the equatorial zone. In Macrobius manuscripts of the Dark Ages we accordingly find simple coastlines of the oikumene incorporating this belt of water.[28] Six-zone maps are associated either with this work of Macrobius or with Martianus Capella (*fl.* 410–439), *De nuptiis Mercurii et Philologiae*. A typical map of this type has lines giving zones with the following widths from north to south: 36°, 30°, 24 + 24° on either side of the equator, 30°, 36°; the 48° belt represents water.

It is unfortunate that, rather than a detailed map, what has survived from early eighth-century Ravenna is a haphazard selection of place-names with lists of tribes corresponding to orientation.[29] Soon after AD 700 a cleric of Ravenna compiled this text to interest his friend the cleric Odo. The date has sometimes been put earlier, but mention of Athanaridus' *Francorum patria* has been taken to point to the early eighth century, with possible later additions.

The preface outlines the geography of the world, referring to the Bible and to Christian cosmology. The nations of the world, particularly the more distant ones, are grouped under twelve night hours, extending from India via Africa to southern Ireland, and twelve day hours, extending from central Ireland via Scanza island (Scandinavia) and Scythia to India. This clockwise enumeration rather suggests that the map followed, or one of those followed, in the listing of tribes was circular or elliptical. The Cosmographer makes it clear that he is only an armchair geographer, an attitude perhaps typical of the period: 'Though I was not born in India nor brought up in Scotia [Ireland], though I did not walk through

Mauretania nor even visit Scythia, I have imbued the whole world by intellectual learning'.

Attempts have been made to reconstruct what is called the Ravenna Cosmographer's world map. Thus Miller has a circular map, with Ravenna off-centre;[30] while an alternative suggestion for the centre is Jerusalem, one which is very common in medieval Christian world maps. But in the first place we cannot be at all sure that the Cosmographer drew any maps, while it seems clear that he possessed and used more than one. If he had, for example, two principal world sources, these may well have been of different periods and have had quite different centres. The chronology is of importance where tribal locations are concerned, since cartographers, if they were conscientious, took account of migrations and moved tribal names accordingly. Secondly, since we know, from the Cosmographer's regional place-names, that he relied heavily on sources that originated under the Roman Empire, we may suspect that his centre was based on a map of that period, possibly Marinus' world map, in which case it is likely to have been Rhodes. Such a centre would suit the opposite pairs of remote areas rather better than Ravenna or Jerusalem. It has to be admitted, however, that neither Rhodes nor Jerusalem is given any topographical prominence. From time to time in the preface the Ravenna Cosmographer, as he is normally known, mentions one or other textual source that he has followed; and modern scholars have suggested some additional ones, such as Pliny and Ptolemy (the latter especially in Asia), which may not have been mentioned because they were not used at first hand. We cannot with confidence build up any theories based on Castorius, several times mentioned as a source by the Cosmographer, since of this fourth-century AD geographer virtually nothing is now known. Miller seems to have been wrong in identifying this Castorius with the compiler of the Peutinger Table.[31] The names in the Cosmography cover far more places than appear in the Peutinger Table: we find tribal names for *civitas* capitals, the addition of the word *colonia* where appropriate, details of military units and mention of rivers, all of which underline the differences between the two.

There are over five thousand place-names listed, with separate sections for rivers and islands; though quite often river names are incorrectly entered as if they were places. The Cosmographer avoids expanding from this, saying 'We could, with the help of

Christ, have mentioned more accurately the harbours and promontories all over the world and the numbers of miles between the cities themselves' (i.18). Despite the mention of Franks, Burgundians, Saxons, Bulgars and other races prominent in the Dark Ages, the place-names tend to be of the Classical period, and prior to the fall of the Western Empire. What is infuriating is the whimsical way in which at one time a string of names is taken in roughly topographical sequence, then suddenly there is a change, and two or three appear from a different part of the province which is being covered. Furthermore, partly because of the Cosmographer's own mistakes and partly through repeated errors in copying, many forms have become ludicrously corrupted. A typical example is the misreading of Fl. (= *Flumen* or *Fluvius*)[32] on maps as if it were El;[33] so that in the British section we have among the many corrupt forms the following:

Elaviana	=	Fl. Abona	Elete	=	Fl. Eitis
Elconio	=	Fl. Cenio	Eltabo	=	Fl. Tavus
Eltavori or Eltanori	=	Fl. Ta..... (R. Tame, Staffordshire)			

The last three of these have, as if they were names of towns, had their nominatives changed to ablatives or locatives. Nevertheless, when corruptions have been accounted for, the Ravenna Cosmography is a source of valuable information which can supplement that from earlier sources.

Large-scale plans are represented by one remarkable example, which shows that exact draughtsmanship was still possible in the Dark Ages. This is the plan of the monastery of St Gall (St Gallen), Switzerland, sent to Abbot Gozbert about 816–23.[34] It was drawn at Reichenau, and is probably connected with monastery reform proposals of 816–17. The whole complex is in neat rectangular plan, with single lines for walls, and each room or building is carefully labelled, including the areas allotted to animals, mainly outside the main compound, and the hostel, outside it. We note that a good-sized square room adjoining the chapel was to serve as a library and scriptorium, and an even larger building as infirmary, with adjacent quarters for doctor, leeching and infirmary chapel; also that there are various workshops in the kitchen area. The scale was $2\frac{1}{2}$ Carolingian feet for each square of a grid which formed the basis of the plan; this is equivalent to 1:192. Such a plan recalls those drawn up by building surveyors under the Roman Empire. It has the type of scale accuracy and care for detail visible in the Forma Urbis Romae and in plans of baths or burial-places. The monasteries were still at that time great repositories of records, such as would earlier have been kept in public buildings, as well as of literary and ecclesiastical manuscripts of many different periods. Monastery library catalogues, though fragmentary and referring only to scattered times and places, show that maps formed a standard element of their holdings.

ARAB USE OF CLASSICAL CARTOGRAPHY

The adaptation by the Arabs of Ptolemaic co-ordinates has already been considered (p. 156), but other aspects of the legacy of Graeco-Roman cartography in the Arab world are also important. One is the desire for beautiful maps, and this is no doubt what is behind the praise of Caliph al-Ma'mun's map as more 'exquisite' than the world maps of Ptolemy, Marinus and others; and some of this beauty no doubt came from Persian sources.

In Graeco-Roman small-scale cartography, north at the top was the commonest, though by no means the only, orientation; but for Arabic mapping the commonest is with south at the top. True, the early map of Balchi, AD 921, shows the Mediterranean, very schematically, with west at the top, Sicily, Crete and Cyprus appearing as circles of equal size.[35] But the maps of al-Idrisi and others have south at the top. Two suggestions made to account for this are: (a) that south was a sacred direction for Zoroastrians, (b) that from Baghdad and other early cultural centres of Islam one looks south to Mecca. But it has also been thought that dynastic Egypt may already have had this orientation.

An important link between European and Arab cartography lay in the world map[36] and book of al-Idrisi. They were commissioned by the Norman King Roger II of Sicily, and were completed in 1154, the book being in Latin and Arabic.[37] Al-Idrisi has been criticized for replacing the *klimata* by 'a series of meaningless lines, which together with pseudo-meridians serve only to divide the map into 70 sections'.[38] But in copies of al-Idrisi's world map[39] the parallels are curved, giving his concept of *klimata* for most of the northern hemisphere. Other Arab cartographers present seven *klimata* in the form of belts separated by straight equidistant lines. The history of *klimata* is a long and complicated one.[40] They seem, as mentioned above (p. 26), to have been invented by Eudoxus of Cnidos, though Honigmann dismisses Strabo's attribution[41] and thinks that Eratosthenes was the inventor. Eratosthenes evidently spoke of *klimata* in the sense of selected latitudes named after particular places and based on the length of the longest day.[42] As suggested by Diller, he may have started with Meroe, Alexandria, Lysimachia (Hellespont) and the R. Borysthenes, having intervals between them of 10,000, 8,000 and 5,000 stades respectively, and with 13, 14, 15 and 16 hours' daylight at the summer solstice, and then have inserted places for 13½, 14½ and 15½ hours somewhat haphazardly. Hipparchus amended Eratosthenes' scheme to provide more scientifically for the decrease in distance towards north. Pliny speaks of parallels, circles or divisions rather than *klimata*:[43] these divisions are seven belts devised, according to him, by 'the ancients', though they differ from those of Eratosthenes, and three more further north. Ptolemy criticizes Marinus' latitudes (as well as longitudes), but in this connection speaks only in terms of particular points of reference, e.g. 'the parallel through Meroe'. Of these Ptolemaic parallels there are in the *Almagest* twenty-six (but only as far north as central Britain; the total would be thirty), in the *Geography* twenty-one.

MAPPAEMUNDI

The term *mappaemundi* (first half from Latin *mappa*, 'table-cloth' or 'napkin', hence originally map on cloth) is used particularly to denote Dark Age and medieval world maps which, mainly Christian, aimed more at expounding a particular concept of the earth than its precise topography.[44] Especially in the later Middle Ages, many large and detailed specimens of them were produced, and of these a small number survive. Modern historians of cartography have approached them with more understanding than critical researchers of an earlier period, viewing them as teaching pictures of the created world.

The commonest shape is circular or oval, the commonest orientation north or east. The centre of the map is often Jerusalem, a choice favoured by Adamnan, *De locis sanctis*; other centres are Mount Sinai and Rome. Certain other places connected with Christianity, such as Antioch and Patmos, or with Classical civilization, such as Constantinople and Mount Olympus, may also be stressed. The Red Sea is not infrequently painted red. Places of historical interest may have brief historical notes attached to them, and indeed the Hereford World Map has a sentence referring to it as *cest estoire*, 'this history'.[45] Scale is erratic, as when Matthew Paris on his map of Britain comments: 'If the page allowed it, this whole island should be longer' (*si pagina pateretur, hec totalis insula longior esse deberet*). In the Jerome map of Asia, Asia Minor as a cradle of Christianity is exaggerated in size. There is usually a *horror vacui*, which results among other things in the drawing in of specimens of exotic or monstrous races such as recall those brought in from Pliny's *Natural History* and from Solinus.

Of the three large late twelfth- or early thirteenth-century Mappaemundi, namely the Vercelli, Ebstorf (destroyed in 1943) and Hereford, it is the last-named, a large map measuring 1 m 63 × 1 m 37, which gives the most information, and much of it reflects Classical sources (pl. 31).[46]

The designer of the map is named as Richard of Haldingham and Lafford, i.e. Holdingham and Sleaford. He is presumed to be the Richardus de Bello who was Prebend of Lafford in Lincoln Cathedral *c.* 1276–83. One of the sources, in wording surrounding the map, is declared as being Orosius: 'Descriptio Orosii de ornesta[47] mundi sicut interius ostenditur'. Another inscription, also surrounding the circular map, refers by name to the Greek cartographers employed by Julius Caesar (p. 40) to prepare a world map. There is also the Biblical sentence[48] *exiit edictum ab Augusto Cesare ut describeretur (h)universus orbis*, which appears to have been inserted in the mistaken belief that Augustus' decree which affected Joseph and Mary was not a registration but a map. At the top is the figure of Christ with angels at the Last Judgment. The earth is circular, with Jerusalem at the centre, and east is at the top of the sheet. The four main cardinal points are indicated by dwarfs, and the remaining eight winds[49] by animals. The basic plan is that of the T-O map, but without the strictly regular divisions of such maps. The names of Paris, Rome and Antioch stand out in detail (even though none is as big as the tower of Babel). This and the marking of two Alpine passes suggest that one object of the Hereford Map was to act as a help to intending pilgrims making their way to Rome, Jerusalem and other places of pilgrimage. Pictorial elements include scenes from the Bible, monstrous races, mythical animals and what looks like a monkey within Norway.

The coastlines even of European countries are very inexact, let alone those of more distant areas. The Red Sea is coloured red, and is much too far from the Mediterranean. With such haphazard mapwork, pictorially attractive as it is, the theory that the Hereford Map was directly influenced by Roman itineraries[50] is difficult to prove, as obviously theologians have imposed their own concepts and technical capability was much reduced as compared with earlier times.

There were also specific medieval itineraries for pilgrims, and of these Matthew Paris (d.1259) in his *Chronica majora*[51] gives what have been called strip maps of pilgrim routes in England and from London to the Holy Land via Sicily. This latter lacks the exactitude of the Bordeaux–Jerusalem itinerary, but attracts the reader by its pictorial elements, including a camel, which might be regarded as extra information about transport in Palestine. Rome, as the chief intermediary place of pilgrimage, has special prominence in size, with an attempt at recognizable illustration (pl. 32).

The rebirth of European cartography exploded in the fourteenth century. It seems to be connected with a new approach to sea charts, and the question arises whether it was inspired by the invention of the magnetic compass, or merely the result of better ships, more trade and information and the resultant wealth used to employ cartographers. As background to the compass, the use of the lodestone, magnetic oxide of iron, is attested from many parts of the world from very early times, and has by some been associated with the Vikings. One of the more specific references is in the English scientific writer Alexander Neckam, *De naturis rerum* (*c.* 1180), to the effect that if sailors lose their bearings they touch a magnetized needle which turns round till it stops with its point facing north. Later tradition assigns its discovery for navigational purposes (perhaps in reality its association with a thirty-two-point windrose) to an unnamed inventor at Amalfi, *c.* 1300. This is also the approximate date of the 'carte Pisane', now in the Bibliothèque Nationale, Paris, covering the Mediterranean and some areas to the north of it. It is the earliest extant portolan map, based on a network of loxodromic lines.[52] Unlike subsequent portolans, it has, in addition to these, two areas of large and four areas of small squares. It is the first map to show the Mediterranean in reasonably correct proportions, and similarly the Black Sea. England is drawn in very poor outline without Scotland but with a few place-names, including a corrupt form of St Thomas of Canterbury. There is no dependence at all on Ptolemy. In Petrus Vesconte's world map of *c.* 1320 the circular form of medieval Mappaemundi is still present, but a sixteen-point windrose is drawn over it, each point being connected by a varying number of lines to other points.[53] As compared with earlier Mappaemundi it has a generally more recognizable shape. Petrus Vesconte also drew regional maps of about the same date to illustrate Marino Sanudo, *Liber secretorum fidelium crucis.*

The Catalan Atlas of 1375 by Abraham Cresques, now in the Bibliothèque Nationale, Paris, is drawn on six wooden tablets, in all 3m wide and 61 cm high.[54] It is not only very artistically painted, with large numbers of human and animal figures, but uses contemporary knowledge to improve the delineation of coastlines in Europe, define the Blue and White Nile, make India far more recognizable, and even incorporate Chinese place-names from records of Marco Polo's travels. Sheets 5 and 6 extend the portolan system to the map of the known world. It has been shown by a distribution grid that it was reasonably accurate in the better-known parts of the world,[55] perhaps the first that owes nothing to Classical antecedents.[56]

Nevertheless, there were Classical revivals. Mention may be made of a world map, of *c.* 1414, now in the Vatican, by Pirrus de Noha,[57] which was drawn to accompany a map of Pomponius Mela. There is no evidence that Mela had maps with his text, but a number of maps of later date are found in his manuscript. Here the cartographer takes some outlines, especially those of southern areas of the world, mainly from Ptolemaic maps, the coastline of India being even more straightened than usual; while other features, such as the shape of Britain and the coastline of the Baltic, are improved from contemporary knowledge.

* * *

In this book an attempt has been made to cover the development of Graeco-Roman terrestrial mapping over a period of two thousand years or more. The arrangement of chapters is partly chronological, partly thematic, as certain types of mapping flourish at different times. For much of the earlier period one has to draw on evidence which is indirect and fragmentary. Research into natural philosophy in the fourth century BC, including the assumption of the earth's sphericity, led to a more scientific approach to terrestrial mapping, which in the late third century BC found its expression in the work of Eratosthenes in Alexandria. But whereas his understanding of the earth's size and mechanism constituted an advance on predecessors, his cartographic method of fitting *sphragides* (irregular polygons) together left something to be desired, and in the second century BC Hipparchus was able to raise the standard of world cartography by geometrical calculation and by evidence from the explorer Pytheas.

The greatest Roman contributions lay in the direction of practical mapping: land surveyors who mapped centuriation, urban surveyors who made the plan of the city of Rome, and road-makers who inspired itineraries and road maps. We may regard Agrippa's map of the known world too as essentially practical, since it attempted to incorporate the latest data on distances available. The skilled work carried out by urban surveyors lasted at least until the early third century AD, whereas world mapping reached its highest point earlier with Marinus and Ptolemy. The latter's sophisticated projection for maps of the known world and his orthogonal regional mapping, with appropriate proportions, are, despite imperfections, vastly superior to anything that came later in Roman and Byzantine cartography. This was followed by medieval *mappaemundi*, which may be explained as products of religious rather than scientific geography. New progress came in the first place with the output of competent portolans promoted by the discovery of the compass, the development of trade, and accurate knowledge of navigation in the Mediterranean and adjacent areas, and later alongside the revival of Ptolemy's *Geography* in the Renaissance. The Arabs had long translated this work and adapted its maps. But this had no appreciable impact in itself on Europeans, and it was left first to the enthusiasm of Planudes (late thirteenth century) in Byzantium and then to Jacopo d'Angelo's translation (1406) of the *Geography* to make the Western world appreciate its true worth.

The continuing competition between European states led to further extension from the old known world, and so to attempts to map Africa, northern Europe and finally the Americas. The cartographers still used the Ptolemaic blueprint, which remained the best basis for world maps and atlas compilation until the modern era.

APPENDIX I

COSMOGRAPHIA IULII CAESARIS

(p. 40)

(A. Riese, ed., *Geographi Latini minores*, pp. 21–3)

In the consulship of Julius Caesar and Mark Antony the whole world was visited under four very wise and chosen men: Nicodemus the east, Didymus the west, Theodotus the north and Polyclitus the south.

From the above date to the consulship of Augustus for the fourth time and Crassus, the east was measured in 21 years 5 months and 9 days. And from the above date to the consulship of Augustus for the seventh time and Agrippa [for the third time] the western part was measured in 26 years 3 months and 17 days. From the above date to the tenth consulship of Augustus the northern part was measured in 29 years 8 months. From the above date to the consulship of Saturninus and Cinna (see below) the southern part was measured in 32 years 1 month and 20 days.

The whole world has 28 seas, 74 islands, 35 mountains, 70 provinces, 264 towns, 52 rivers, 129 tribes.

The east has 8 seas, 8 islands, 7 mountains, 7 provinces, 70 towns, 17 rivers, 46 tribes.

The western part has 8 seas, 17 islands, 9 mountains, 24 provinces, 77 towns, 14 rivers, 29 tribes.

The northern part has 10 seas, 32 islands, 12 mountains, 16 provinces, 61 towns, 16 rivers, 29 tribes.

The southern part has 2 seas, 16 islands, 6 mountains, 23 provinces, 46 towns, 5 rivers, 24 tribes.

NOTES

1 The alleged years of completion are given on p. 40, except for the south. If this refers to the consulship of C. Sentius Saturninus the elder, for which Spanish *fasti consulares* wrongly give the colleague as Cinna, the year is 19 BC; if it is the consulship of C. Sentius Saturninus his son, it is AD 4, though Cn. Cornelius Cinna Magnus was consul not then but in AD 5.

2 Some totals of the four zones given above show slight discrepancies

from the world figures given. But they are unlikely to have been borrowed from what follows. The latter is headed *excerpta eius sphaerae vel continentia*, and is said at the end to have been published without permission by a pupil of Julius Honorius. The items are very muddled, many of the 'towns' being actually names of peoples; the totals differ very considerably from those given above; and their late date is evidenced from the name *Constantina* and probably from tribal movements. One may conclude that Honorius may to some extent have drawn on Julius Caesar's map to construct a globe. It seems unlikely that *eius* refers to Julius Caesar; there may be a lacuna in the text.

APPENDIX II

PLINY, *NATURAL HISTORY* vi.211–20

(*cf. p. 67*)

I shall also include one more Greek invention of extraordinary subtlety, so that my survey of world geography may lack nothing, and so that by indicating the regions one may ascertain what affiliation in relationship of days and nights each region has, and which of them have shadows of equal length and an equivalent curvature of the *mundus* (heaven and earth)*. So this too will be related, and the whole earth will be apportioned to parts of the heaven.

There are many zones of the universe (*segmenta mundi*) which our writers have called little circles (*circuli*), while the Greeks have called them parallels. I will begin with the southern part of India; this zone extends as far as Arabia and the coasts of the Red Sea. Those included are the Gedrosi, Carmani, Persians, Elymaei, Parthyene, Aria, Susiane, Mesopotamia, Babylonian Seleucia, Arabia as far as Petra, Coele Syria, Pelusium, Lower Egypt known as Chora, Alexandria, coastal areas of the province of Africa, all the towns of Cyrenaica, Thapsus, Hadrumetum, Clupea, Carthage, Utica, the two Hippos, Numidia, the two Mauretanias, the Atlantic and the Pillars of Hercules. At these latitudes at midday at the equinox, a 'navel' which they call a gnomon 7 ft long throws a shadow not more than 4 ft long. The longest period of night or day contains 14 equinoctial hours, the shortest 10.

The second *circulus* begins in western India and goes through central Parthia, Persepolis, the nearest parts of Persis, the nearer part of Arabia, Judaea and the Mt Lebanon area, and embraces Babylon, Idumaea, Samaria, Jerusalem, Ascalon, Joppa, Caesarea, Phoenicia, Ptolemais, Sidon, Tyre, Berytus, Botrys, Tripolis, Byblos, Antioch, Laodicea,

*The Latin is *quibusque inter se pares umbrae et aequa mundi convexitas*. The words in Rackham's translation, 'the world's convexity is equal', are too literal and do not begin to interpret. The phrase in the next sentence, *terraque universa in membra caeli digeretur*, show that Pliny conceived of a spherical universe in which the earth and the heavens are inextricably linked. From the point of view of geocartography the effect is the same, referring to equal length of day and night round a certain zone of the earth's circumference.

Seleucia, the coastal area of Cilicia, the south of Cyprus, Crete, Libybaeum in Sicily and the northern parts of the provinces of Africa and of Numidia. A 35-ft gnomon at the equinox throws a shadow 24 ft long. The longest day and night are 14.4 equinoctial hours.

The third *circulus* starts with India closest to the Imavus (Himalayas). It goes through the Caspian Gates, the nearest parts of Media, Cataonia, Cappadocia, the Taurus, Amanus, Issus, the Cilician Gates, Soli, Tarsus, Cyprus, Pisidia, Pamphylia, Side, Lycaonia, Lycia, Patara, Xanthus, Caunus, Rhodes, Cos, Halicarnassus, Cnidos, Doris, Chios, Delos, the central Cyclades, Gythion, Malea, Argos, Laconia, Elis, Olympia, Messenia in the Peloponnese, Syracuse, Catina, central Sicily, southern Sardinia, Carteia, Gades. A gnomon 100 in. long throws a shadow 77 in. long. The longest day is $14.5\dot{3}$ equinoctial hours.

The fourth *circulus* has lying under it the areas on the other side of the Imavus, the south [*sic*] of Cappadocia, Galatia, Mysia, Sardis, Smyrna, Mt Sipylus, Mt Tmolus, Lydia, Caria, Ionia, Tralles, Colophon, Ephesus, Miletus, Chios [cf. third *circulus*], Samos, the Icarian Sea, the northern Cyclades, Athens, Megara, Corinth, Sicyon, Achaia, Patrae, the Isthmus, Epirus, the north of Sicily, the east of Gallia Narbonensis, the coast of Spain from Nova Carthago and the areas west of it. To a gnomon 21 ft long correspond shadows 16 ft long. The longest day has $14.\dot{6}$ equinoctial hours.

The fifth segment comprises, from the entrance of the Caspian, Bactria, Hiberia, Armenia, Mysia, Phrygia [cf. fourth *circulus*], the Hellespont, the Troad, Tenedos, Abydos, Scepsis, Ilium (Troy), Mt Ida, Cyzicus, Lampsacus, Sinope, Amisus, Heraclea Pontica, Paphlagonia, Lemnos, Imbros, Thasos, Cassandria, Thessaly, Macedonia, Larisa, Amphipolis, Thessalonica, Pella, Edesus, Beroea, Pharsalia, Carystos, Boeotian Euboea, Chalcis, Delphi, Acarnania, Aetolia, Apollonia, Brundisium, Tarentum, Thurii, Locri, Regium, the Lucani, Neapolis, Puteoli, the Tuscan Sea, Corsica, the Balearic Islands and central Spain. 7-ft gnomon: 6-ft shadow. The longest day has 15 equinoctial hours.

The sixth group, which includes the city of Rome, embraces the Caspian tribes, the Caucasus, the north of Armenia, Apollonia on the Rhyndacus, Nicomedia, Nicaea, Chalcedon, Byzantium, Lysimachia, the Chersonese, the Black Bay (Melas sinus), Abdera, Samothrace, Maronea, Aenos, the area of the Bessi, Thrace, the area of the Maedi, Paeonia, Illyria, Durrachium, Canusium, the fringe of Apulia, Campania, Etruria, Pisae, Luna, Luca, Genua, Liguria, Antipolis, Massilia, Narbo, Tarraco, central Hispania Tarraconensis and from there through Lusitania, 9-ft gnomon: 8-ft shadow. The longest day is $15.\dot{1}$ hours or according to Nigidius 15.2.

The seventh division begins on the other side of the Caspian and goes above Callatis, the (Cimmerian) Bosporus, the R. Borysthenes, Tomi, the further parts of Thrace, the Triballi, the rest of Illyria, the Adriatic, Aquileia, Altinum, Venetia, Vicetia, Patavium, Verona, Cremona, Ravenna, Ancona, Picenum, the Marsi, the Paeligni, the Sabines, Umbria, Ariminum, Bononia, Placentia, Mediolanum and all areas from the Apennines and across the Alps to Aquitanian Gaul, Vienna (Vienne), the Pyrenees and Celtiberia. 35-ft 'navel': 36-ft shadows, except that in part of Venetia the shadow is equal to the gnomon. The longest day is 15.6 equinoctial hours.

Up to now we have published the findings of ancient writers. The most careful of subsequent writers have assigned to segments the rest of the earth: from the R. Tanais through Lake Maeotis and the Sarmatae to the R. Borysthenes and so through the Dacians and part of Germany to the Gallic provinces, embracing the shores of the ocean, to form a day-length of 16 hours. A second, through the Hyperboreans and Britain, has a day-length of 17 hours. Finally a Scythian segment, from the Rhipaean ranges to Tyle (Thule), where there are periods of continuous day and continuous night. The same writers also place two *circuli* before our starting point: the first through the island of Meroë and Ptolemais on the Red Sea (built for elephant hunting), where the longest day is 12.5 hours, the second going through Syene in Egypt, with a 13-hour day. The same writers have added half an hour to each of the *circuli* up to the last.

APPENDIX III

DATA ON THE ORANGE CADASTERS

(pp. 108–10)

MATERIAL FOUND

1 The inscription recording Vespasian's edict.
2 Fragments of cadasters A–C. Many fragments were virtually destroyed through a collapse at the Orange museum in 1962; the remainder are in the museum. Exhibits include a rectangle of Cadaster B 25⅓ 'centuries' high and 39 'centuries' wide.
3 Fragments from the public record office.

ABBREVIATIONS

1 TRIC RED (Cadaster B), *Tricastinis reddita*, 'restored to the Tricastini'. The lands of this local tribe were at first confiscated, but it seems that under the Flavian emperors much of what was in effect poorer land was given back to them.
2 EXTR or EXT, *ex tributario*, 'formerly tribute-paying'. The Tricastini had to pay tribute, the legionary veterans and their heirs did not.
3 REL COL or REL or COL, *reliqua coloniae*, 'remaining in possession of (or remaining to) the colony'. Lands not allocated to veterans belonged to the community and could, for example, be let out as pasture.
4 R P, *rei publicae*, 'State lands' (Cadaster A). As this is not found in B and C, it must represent a category which by the later date of these had become merged with No. 3.
5 SVBS, *subseciva*, 'incomplete lots'. These might either be land unsuitable for centuriating or land between 'centuries' and the outer boundary of the territory.
6 IVG, *iugerum*, 28,800 square feet (Roman), *c.* 0.623 acres/0.252 ha.; S = ½ iugerum; £ = 1200 square feet.
7 DD, SD, VK, CK: see p. 90.

SYMBOLS DENOTING RENTS

Annual rents are quoted in *denarii* and *asses*. Originally there were 10 *asses* to a *denarius*, but in the 140s BC the *as* was re-tariffed at 16 to the *denarius*.

Other units are as follows:

libella = $\frac{1}{10}$ *denarius* *singula* = $\frac{1}{2}$ *libella* *teruncius* = $\frac{1}{2}$ *singula*

The symbols, in the order used, are:

X *1 denarius*	= *2 libellae*	T *1 teruncius*
S $\frac{1}{2}$ *denarius (semis)*	≡ *3 libellae*	AI *1 as*
— *1 libella*	£ *1 singula*	

Thus 11 *asses* are represented by S-TAI, i.e.

$$\frac{1}{2} + \frac{1}{10} + \frac{1}{40} + \frac{1}{16} = \frac{11}{16} \textit{ denarius.}$$

SPECIMEN 'CENTURIES'

Fig. 30 shows parts of two 'centuries', one on each side of the *kardo maximus*: Block IIIC, no. 171 Piganiol.

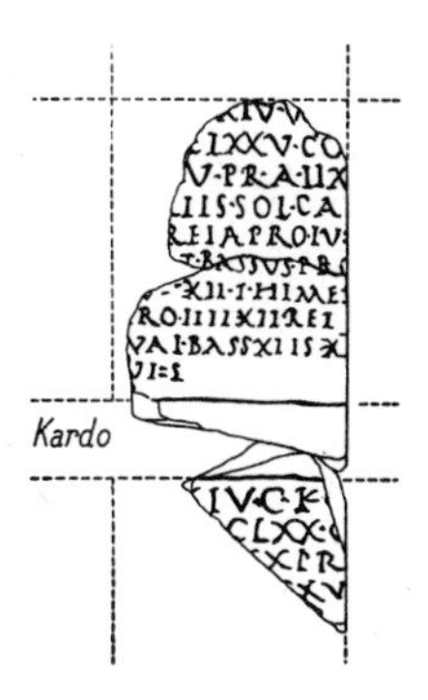

Fig. 30. Arausio (Orange), Cadaster B, Group IIIC.

sinistra decumani XIV ultra kardinem I: ex tributario CLXXV (iugera), coloniae XXV (iugera), praestant aera IIX, denariis XIIS; solvit Careia pro IV (iugeribus) et quadrante denarios II libellam et teruncium, Bassus pro IX (iugeribus) et quadrante denarios II libellam et teruncium, item et pro IIII (iugeribus) denarios II; reliqua Valerio Basso XII (iugera) et semis, denarios sex, duas libellas et singulam.

(KM)

sinistra decumani XIV citra kardinem I: ex tributario CLXX (iugera), coloniae XXX (iugera), praestant aera IIX, denariis XV; solvit. . . .

Translation:

Left of *decumanus* 14, beyond *kardo* 1. Withdrawn from tribute-paying lands, 175 *iugera*. In possession of colony, 25 *iugera*. Tariff 8 *asses* per *iugerum*, total rent 12$\frac{1}{2}$ *denarii*. Rents payable: Careia, for 4$\frac{1}{4}$ *iugera*, 2 *denarii* 2 *asses*. Bassus, for 4$\frac{1}{4}$ *iugera*, 2 *denarii* 2 *asses*. Bassus, for 4 *iugera*, 2 *denarii*. Remainder to Valerius Bassus, 12$\frac{1}{2}$ *iugera*, 6 *denarii* 4 *asses*.

Left of *decumanus* 14, this side of *kardo* 1. Withdrawn from tribute-paying lands, 170 *iugera*. In possession of colony, 30 *iugera*. Tariff 8 asses per *iugerum*, total rent 15 *denarii*. Rents payable. . . .

APPENDIX IV

PTOLEMY, *GEOGRAPHY* ii.3, Manuscript variants on British place-names

X = Vaticanus graecus 191
S (Σ) = Florentinus Laurentianus xxviii.9.
The symbol j is used for S (Σ) and kindred manuscripts, the apparatus criticus being purposely simplified; see Müller's edition, vol. I.i.

	Longitude E. of Canaries		*Latitude*		*MSS*
(a) Capes, bays and rivers					
Longos R.	24	30	60	40	j
	24	00	60	40	most others
Abravannos R.	19	20	60	15	j (X corrupt)
	19	20	61	00	most others
Iena estuary	19	00	60	20	j
	19	00	60	30	most others
Promontory of the Ganganoi	15	50	56	50	Xj
(Kaiankanoi X)	15	00	56	00	other MSS
Stukkia R.	15	20	55	50	Xj
	15	20	55	30	other MSS
Tuerobis R.	15	30	55	10	j
	15	00	55	00	other MSS
Promontory of Hercules	14	00	52	45	Xj
	14	00	53	00	most others
Antivestaion or Bolerion	11	30	52	30	X
promontory	11	00	52	30	other MSS
Iska R.	17	40	52	20	Xj
	17	00	52	20	other MSS
Alaunos (1) R.	17	20	52	20	j
	17	40	52	40	other MSS
Virvedron promontory	31	20	59	40	j
	31	00	60	00	most others
Loxa R.	27	30	59	20	j
	27	30	59	40	other MSS
Kailis R.	27	20	58	45	j
	27	00	58	45	other MSS

	Longitude E. of Canaries		*Latitude*		*MSS*
Tava estuary	25	00	58	50	j
	25	00	58	30	other MSS
Boderia estuary	22	30	59	00	j, some others
	22	30	58	45	other MSS
Alaunos (2) R.	21	20	58	30	j
	21	40	58	30	other MSS
Dunon bay	20	30	57	30	X
	20	15	57	30	most others
Bay of the Gabrantovices	21	00	57	30	j
	21	00	57	00	other MSS
Abos R.	21	00	56	40	X
	21	00	56	30	other MSS
Gariennos R.	20	50	55	40	j
	21	00	55	20	other MSS
Sidumanis R. (var.	20	30	55	00	Xj
Eidumanios)	20	10	55	00	most others
Iamesa (= Tamesa)	20	30	54	45	X
estuary	20	30	54	30	other MSS
Kantion promontory	22	30	54	00	X
	22	00	54	00	other MSS

(b) Places

Karbantorigon	19	00	59	45	j
	19	00	59	30	other MSS
Kolani(k)a	20	45	59	00	Xj
	20	20	59	10	most others
Alauna	22	45	59	50	j, some others
	22	45	59	20	other MSS
Devana	26	15	59	00	Xj, some others
	26	00	59	45	other MSS
Vinnovion (see	17	45	58	00	j
below)	17	30	58	00	other MSS
Brannogenion	16	45	56	15	j
	16	00	56	15	other MSS
Virokonion	16	45	55	10	Xj
	16	45	55	45	other MSS
Salinai	20	45	55	50	j
	20	45	55	20	some others
	20	10	55	40	some others
Kamulodanon	21	15	55	00	Xj
(= Camulodunum)	21	00	55	00	other MSS
Noiomagos	19	45	53	10	j
	19	45	53	25	some others
	19	45	53	20	some others
Iskalis	16	00	53	40	Xj, some others
	16	00	53	30	most others

	Longitude E. of Canaries		*Latitude*		*MSS*
Venta [Belg.]	18	40	53	00	Xj
	18	40	53	30	most others
Dunion	18	00	52	40	Xj, some others
	18	00	52	55	most others
Voliba	14	45	52	00	j, some others
	14	45	52	20	most others
Tamare	15	30	52	45	Xj
	15	00	52	15	other MSS
Iska [Dumnon.]	17	30	52	50	X
	17	00	52	50	j
	17	30	52	45	most others

These variants are sometimes due to rounding off or copyist's omissions, sometimes intentional. The most serious misplacing among the names listed is Vinnovion (Binchester), which is very far east of either variant.

APPENDIX V

PEUTINGER TABLE

Categories of places

Personifications (3):
Roma, Constantinopolis, Antiochia.

Polygonal walled cities (6): Aquileia, Ravenna, Tessalonice, Nicomedia, Nicea, [Ancyra].

Baths and seaside resorts (48): for subdivisions of characteristic sign see Annalina and M. Levi, *Itineraria picta* (Rome, 1967), pp. 222–3. Excluded from the list below are their subdivisions 12, 19 and 21, which have lozenge patterns and do not seem to belong to this category.

Segment (K. Miller)	*Name in Peutinger Table*	*Name in* RE
II.1	Aquis	Aquae Convenarum
II.2	Praetorium Agrippine	Praetorium Agrippinae
II.4	Aquis Nisincii	Aquae Nisinciae
II.4	Aquis Calidis (1)	Aquae Calidae
II.4	Aquis Segeste	Aquae Segestae
II.4	Aquis Bormonis	Aquae Bormonis
II.5	Aquis Segete	Aquis Segete
II.5	Indesina	Indesina
III.1	Aquis Sestis	Aquae Sextiae
III.4	Aquis Tatelis	Aquae Statiellae
IV.1	Ad aquas Herculis	ad Aquas Herculis
IV.1	Aquis Thibilitanis	Aquae Thibilitanae
IV.1	Ad taberna frigida	Taberna frigida
IV.2	Aquae Populaniae	Aquae Populoniae
IV.2	Aquas Volaternas	Aquae Volaterranae
IV.4	Ad aquas casaris	Aquae Caesaris
IV.5	Ad aquas	ad Aquas
IV.5	Fonte Timavi	Fons Timavi

Segment (K. Miller)	*Name in Peutinger Table*	*Name in* RE
V.1	Aquas passaris	Aquae Passeris
V.1	Quaeri	—
V.2	Mindo fl.	Minio (river in Etruria)
V.3	Aquis	Aquae
V.3	Vacanas	Baccanae
V.3	Aquas tauri	Aquae Tauri
V.3	Aquas Apollinaris	Aquae Apollinares
V.4	Aquae cutillie	Aquae Cutiliae
V.5	Ad pretorium	Praetorium
VI.1	Ad pretorum	Praetorium
VI.1	Tres tabernas	Tres tabernae
VI.1	Servitio	Servitium
VI.1	Ad aquas	ad Aquas
VI.2	—	(Appii Forum)
VI.3	Ad nonum	—
VI.3	Invinias	in Vinias
VI.3	Syllas	Syllae
VI.3	—	(Baiae)
VI.3	Pretorium . . . Nucerie . . .	Luceria
VI.3	Siclis	Siclis
VI.3	Ad aquas	Aquae Tacapitanae
VI.4	Inaronia	Aronia
VI.5	Oplont–s	Oplontis
VI.5	Ad teglanum	ad Teglanum
VII.1	Stanecli	Stanecli
VII.1	Aquas labodes	Aquae Labodes
VII.1	Aque Ange	Aquae Angae
VII.2	—	(Columna Regia)
VII.5	Ad aquas	ad Aquas
VIII.4	Aquis calidis	Aquae Calidae
X.2	Aquis calidis	—
XI.2	—	—

Granaries etc. (10)

V.2	Centum cellis	Centum cellae
V.4	Aqua viva	Aqua viva
VI.1	Bobellas	Bovillae
VI.1	Sublanubio	Sublanuvio

Segment (K. Miller)	*Name in Peutinger Table*	*Name in* RE
VI.3	Epetio	Epetium
VII.3	—	(Vax)
IX.1	Chrisoppolis	Chrysopolis
IX.2	Livissa	Libyssa
IX.2	Eribulo	Eribolon
X.4	—	(Pictanus)

For *ad horrea* (Horrea Caelia, VI.4) a combination of town and granary symbols is used.

Temples etc.

(a) *Religious connotations*: Iseum (3); Serapeum (3); ad Dianam (6); Templo Iovis, Ins. Iovis, iovisurius (*RE* s.v. Bosporus 92), Iovis penninus id est Agubio (= Iuppiter Apenninus, Iguvium), Iovis tifatinus; Templum Minervae; Templ. Veneris; Templ. Herculis, Ad herculem; Ad mercurium; Templ. Augusti; Fano Fugitivi; Fano Furtune (Fanum Fortunae); Balacris (Balagrai), hoc est templum Asclepy.

(b) *Other names*: Durocortoro; Amurio (Amorion); Saldas Colonia; Aventicum Hel(v)etiorum; Ivavo (= Iuvavum); NS. cephalania (Kephallenia); NS. Achillis; Sycas (Sykai); Augusta Ruracum (Augusta Rauricorum); Fons co Neapolis (*RE* s.v. Neapolis 11); Cabillione (Cabillonum); Ionnaria; Sestos.

(c) 4 unnamed places.

Harbours (2): [Portus, N. of Ostia], Fossis Marianis

Lighthouses (2): [Bosphorus], [Alexandria]

Altars (3): Arephilenorum (Arae Philaenorum) viii. 2; Ara Alexandri XII.3; Hic Alexander responsum accepit Usque quo Alexander XII. 5.

Individual buildings (3): Ad matricem VI.5; Vadis Sobates (Vada Sabatia) III.4; Tragurio VI.3.

Tunnel (1): [Crypta Neapolitana] VI.4.

Cities and towns. For list and possible subdivisions see Annalina and M. Levi, *Itineraria picta* (Rome, 1967), pp. 215 ff. The fact that certain variants are found only in a limited part of the map may point to the whims of individual cartographers rather than to any systematic differentiation.

APPENDIX VI

GREEK AND ROMAN WORDS FOR 'MAP'

Greek

1. γῆς περίοδος, 'map of the earth', lit. 'way round the earth': Hdt. iv.36, v.49; Ar. *Nub.* 206; Arist. *Mete.* 362b12; Agathem. i.1. Kindred sense: 'book of geography': Arist. *Pol.* 1262a19; *Rhet.* 1360a34; *Mete.* 350a16. Note: περίοδος alone can mean 'going round', 'circumference', 'period', 'orbit', 'region' (inscription from Cos). Also γῆς καὶ θαλάσσης περίμετρον, Diog. Laert. ii. 2. 1.

2. πίναξ, 'map', lit. 'board': Hdt. v.49: Plut *Thes.* 1; πίναξ γεωγραφικός, Strabo i.1.11. Cf. πινακογραφία, 'drawing of maps', Strabo ii.1.11; πινακογράφος, Eust. *ad D.P.* 4. Kindred senses: 'astronomical table', 'picture', 'drawing or writing tablet', 'catalogue', 'archive' (inscription from Delphi, second century BC).

3. σφαῖρα, 'globe', 'sphere' of any kind.

4. σχηματογραφία, 'plan or map of land': *P. Meyer* i.20 (second century BC): *PSI* 10.1118.10 (first century AD). Cf. σχηματογραφέω, 'draw a plan'.

5. ἰχνογραφία, 'ground-plan': Vitr. i.2.2.

6. διάγραμμα, 'drawing', 'diagram'; 'map outline': Julian, *Epist.* 7 (403 C-D).

Latin

1. *forma*, 'map', 'plan', lit. 'shape': 'Lex Thoria' of 111 BC, Bruns *FIR* i.86. § 78; Cic. *fam.* ii.8.1; *Q.fr.* ii.5.3 (ii.6.2); Arausio, inscription of Vespasian; Suet. *Caes.* 31. See also *Thesaurus Linguae Latinae* s.v. forma III.B.1.c; *Corpus agrimensorum* ed. Blume et al., Index verborum s.v. forma. Kindred senses: 'shape' (including countries, hence liable to confusion: see R. K. Sherk, 'Roman geographical exploration and military maps', *ANRW* ii.1 (1974), 534–62; 'geometrical figure': Livy xxv.31.9; Quint. i.10.35.

2. *tabula* 'map', lit. 'tablet' (cf. πίναξ): Cic. *Att.* vi.2.3 (of Dicaearchus); Prop. iv.3.37. Kindred senses: 'public record', 'writing tablet', 'picture', 'plot of ground'.

3. *descriptio*, 'world map', lit. 'drawing': Vitr. viii.2. Kindred senses: 'drawing', 'copy', 'description'.

4. *sphaera*: see σφαῖρα.

5. *itinerarium pictum*: Veget. iii.6. It is uncertain whether the phrase refers to a road map of the Peutinger Table type or to an itinerary accompanied by paintings.

6. *mappa*, late Latin, 'map', lit. 'cloth': *Grom.* p. 358, 12 *mappa quas lineas habuerit observetur*; Ratpert. *Cas. S. Galli* 10, *SS.*, II p. 72 *unam mappam mundi*. Kindred sense: 'allotment'.

7. *charta*, Renaissance Latin, 'map', 'chart' (*c.* 1550), lit. 'papyrus'; use extended in medieval Latin to any document on papyrus, parchment, vellum, etc.

APPENDIX VII

PTOLEMY MANUSCRIPTS WITH MAPS

Greek maps

(a) Redaction A (27 maps in origin)

Symbol	*Name*	*Date*	*Maps*
F	Fabricianus bibl. univ. Havniensis graecus 23, Copenhagen	13th cent.	Fragmentary; originally world + 26 regional (only 3 extant)
U (Ur)	Vaticanus Urbinas graecus 82	Late 13th cent.	World + 26 regional
K	Constantinopolitanus Seragliensis 57, Istanbul	Late 13th cent.	World + 26 regional (poorly preserved)
L	Vatopedi 655, Mt Athos + British Library Add. 19391 + some pages in Leningrad	Early 14th cent.	World + 26 regional (world map + British Isles in London, remainder except for 2 half maps at Vatopedi)
—	Seragliensis graecus 27, Istanbul	15th cent.	World + 4 'continental' + 26 regional
R	Marcianus graecus 516, Venice	15th cent.	Originally world + 26 regiona (world map, 2 others and 2 hal maps missing)
—	Laurentianus Soppr. 626, Florence	15th cent.	World + 4 unfinished maps
—	Parisinus graecus 1402	15th cent.	World + 4 unfinished maps
—	Marcianus graecus 388, Venice	15th cent.	World + 26 regional
—	Historicus graecus 1, Vienna, Österreichische National-bibliothek	1454	World + 26 regional
—	Vossianus graecus F.1, Leiden, University Library	16th cent.	No text; world + 26 regional

•) Redaction B (64 or 68 maps in origin)

ymbol	Name	Date	Maps
(Ω)	Laurentianus Plut. XXVIII.49, Florence	Early 14th cent.	Originally world + 1 Europe, 2 Asia, 2 Africa, 63 regional (total 69, of which 65 extant)
	Burney 111, British Library	14th cent.	Maps derived from O
	Constantinopolitanus chartaceus, Istanbul	14th–15th cent.	69 maps as originally in O
	Palatinus graecus 388, Vatican	1400–1450	World + 63 regional
	Mediolanensis graecus D 527, Ambrosiana, Milan	1400–1450	World + 63 regional

Arabic maps

(a) Ptolemy
Constantinopolitanus Arabicus 2610, Aya Sofia, Istanbul: World + 1 Europe, 2 Asia, 1 Africa + Corsica, Peloponnese and Crete from Redaction B, others from Redaction A

(b) The adaptation by al-Khwarizmi
Strasbourg, Bibliothèque Universitaire et Régionale, L. arab. cod. Spitta 18: map of the Nile

Latin maps of Redaction A (numbers as in J. Fischer)

(a) Palla Strozzi, Lapaccino, Boninsegni and copies from these editors:

L1 Vaticanus latinus 5698; L2 Nanceianus latinus 441, Bibliothèque Publique, Nancy; L3 Atlas Laurentianus Aedilium 175, Florence; L4 Parisinus latinus 4801; L5 Neapolitanus latinus V.F.33, Biblioteca Nazionale, Naples; L6 Oratorianus lat. Pil. IX, n.II, Biblioteca dell'Oratorio, Naples; L7 Ambrosianus B52 Inf., Milan; L8 Parisinus latinus 4803; L9 Parisinus latinus 4804; L10 Harleianus latinus 7195, British Library; L11 Harleianus latinus 7182, British Library; L12 Parisinus latinus 15184; L13 Zeitzianus latinus 497, Stiftsgymnasium, Zeitz, E. Germany; L14 Bibliothecae militaris 26843, Public Library, Leningrad; L39 Constantinopolitanus Seragliensis latinus 44, Istanbul.

(b) Donnus Nicolaus Germanus and copies:

L15 Latinus V.F.32, Biblioteca Nazionale, Naples; L16 Ebnerianus latinus, Lenox Library, New York Public Libraries; L16a Parisinus latinus 10764; L17 Valentianus latinus 1895, Biblioteca Universitaria, Valencia; L18 Estensis latinus 463, Biblioteca Estense, Modena; L18a Parmensis latinus 1635, Biblioteca Palatina, Parma; L19 Parisinus latinus 4805; L20

Laurentianus Plut. XXX, 3, Florence; L21 Zamoiskianus latinus, National Library, Warsaw; L22 Urbinas latinus 274, Vatican; L23 Urbinas latinus 275, Vatican; L24 Wolfeggius latinus, Schloss Wolfegg, Württemberg; L24a Monacensis misc. lat. 10691, Bayerische Stadtbibliothek, Munich; L24b Vadianus latinus, Stadtbibliothek, St Gallen, Switzerland; L24c Bruxellensis latinus 7350 (14887), Bibliothèque Royale, Brussels; L24d Monacensis latinus 29, Bayerische Stadtbibliothek, Munich, maps only; L24e Monacensis latinus 388, Bayerische Stadtbibliothek, Munich; L24f Newberriensis latinus, Newberry Library, Chicago; L25 Laurentianus Plut. XXX, 4, Florence; L26 Vaticanus latinus 3811, no maps.

(c) Massaio
L27 Vaticanus latinus 5699; L28 Urbinas latinus 277, Vatican; L29 Parisinus latinus 4802; L30 H. M. 1092, Henry E. Huntington Library and Art Gallery, San Marino, California.

(d) Berlinghieri
L31 Braidensis 26 (AC XIV.44), Biblioteca Nazionale, Milan; L32 Urbinas latinus 273, Vatican; L33 Parisinus latinus 8834; L34 Laurentianus Plut. XXX, 1, Florence.

(e) Henricus Martellus Germanus
L35 Florentinus Magliab. Cl.XIII Cod.16, Biblioteca Nazionale, Florence; L36 Vaticanus latinus 7289.

(f) Others
L37 Laurentianus Plut. XXX, 2, Florence; L38 Escurialensis e.γ.1, El Escorial; — Harleiensis 3686, British Library, London, 18 rough maps (13 of Books I-III, 5 'continental'); — Nordenskiöld Collection Msc.1, University Library, Helsinki; — Bibliotheca Jagellon. 7805, Kraków; — Matritensis Vitr. 12.2 (R.III.23, Res.255), Biblioteca Nacional, Madrid; — Paris, B. Arsénal 8536 (Ital.42); — Salamanca, Biblioteca Universitaria 2586 (Palacio VIII. S.1); L40 Wilczek-Brown Codex, Providence, R.I., Brown University, The John Carter Brown Library (L. Bagrow, *Imago Mundi* XII, 1955, 171–4), no text, 14 maps (Europe 10, Africa 4), preceded by printed Berlinghieri world map; — Codex 161e, Biblioteca Federiciana, Fano.

Latin map of Redaction B

Fondo principale, Biblioteca Nazionale, Naples.

See O. Cuntz, ed., *Die Geographie des Ptolemaeus* (Berlin, 1923); J. Fischer, ed., *Claudii Ptolemaei Geographiae Codex Urbinas Graecus 82* (Leiden, Leipzig and Turin, 1932); M. Destombes, *Mappemondes A.D. 1200–1500*

(Amsterdam, 1964), pp. 247–8; J. Babicz in *The History of Cartography*, ed. B. Harley and D. Woodward, vol. 3 (Chicago and London, forthcoming); D. W. Marshall, 'A list of manuscript editions of Ptolemy's *Geography*', *Bulletin, Geography and Map Division, SLA* (Special Libraries Association) 87 (1972), 17–38.

NOTES

ANRW — *Aufstieg und Niedergang der römischen Welt*
BAR — British Archaeological Reports
CIL — *Corpus Inscriptionum Latinarum*
JHS — *Journal of Hellenic Studies*
JRS — *Journal of Roman Studies*
RE — Pauly-Wissowa, *Realenzyklopädie der klassischen Altertumswissenschaft*
RIB — *Roman Inscriptions of Britain*
RPh — *Revue de Philologie*

Chapter I

1 A. R. Millard in vol. 1 of the forthcoming *History of Cartography*, ed. J. B. Harley and David Woodward (Chicago University Press).
2 P. D. A. Harvey, *The History of Topographical Maps* (London, 1980), pp. 122–3, figs. 68–9.
3 *Ibid.* pp.124–5, figs. 70–71; McG. Gibson, 'Nippur: new perspectives', *Archaeology* (USA) 30 (1977), 26–37, esp. 34–7.
4 S. H. Langdon, 'An ancient Babylonian map', *The Museum Journal* VII (1916), 263–8.
5 A. T. Clay, 'Topographical map of Nippur', *Trans. of the Department of Archaeology, Univ. of Pennsylvania Free Museum of Science and Art* 1/III (1905), 223 ff.
6 A. R. Millard and K. H. Deller in Martha A. Morrison and David I. Owen, ed., *Studies on the Civilization and Culture of Nuzi and the Hurrians* (Winona Lake, 1981), p. 438.
7 Campbell Thompson, *Cuneiform Texts* 22, pl. 48; E. Unger, *Babylon: die heilige Stadt . . .* (repr. Berlin, 1970), pp. 254–8; A. L. Oppenheim in C. C. Gillispie, ed., *Dictionary of Scientific Biography*, vol. 15 (New York, 1978), p. 637.
8 Cf. the Epic of Gilgamesh, transl. N. K. Sandars (rev. edn., Harmondsworth, 1973).
9 O. Neugebauer, *The exact Sciences in Antiquity*, 2nd edn. (Providence, R.I., 1957), p. 35.
10 K. Cebrian, *Geschichte der Kartographie* vol. only published, Gotha, 1922), p. 46.
11 *RE* s.v. Karten (Palästina).
12 A. F. Shore in *History of Cartography* (n. vol. 1.
13 T. E. Peet, *The Rhind mathematical Papy* (Liverpool, 1923); A. B. Chace, *The Rhi mathematic Papyrus*, 2 vols. (Oberlin, 1927–9
14 Turin, Museo Egizio: J. Ball, *Egypt in Classical Geographers* (Cairo, 1942), pp. 180–8 pls. VII, VIII; G. Goyon, 'Le papyrus de Tu dit "des mines d'or" et le Wadi Hammama *Annales du Service des Antiquités de l'Egy* XLIX (1949), 337–92; E. Scamuzzi, *Mu Egizio di Torino* (Turin, 1963), pl. 88; Posener et al., *A Dictionary of Egypti Civilization* (London, 1962), p. 112.
15 Ed. H. Schoene, vol. III (Leipzig: Teubn 1903).
16 Herodotus iii.60.
17 A. Badawy, *A History of Egyptian Archit ture: the Empire (the New Kingdom)* (Berkel and Los Angeles, 1968), pp. 488–99. For plans houses and gardens see J. Wilkinson, *T Manners and Customs of the Ancient Egyptia* (London, 3rd edn., 1847), ii.5, fig. opp. p. 9 105, 129–48; J. Capart, *Egyptian Art: I troductory Studies*, transl. W. R. Daws (London, 1923), pp. 86–91.
18 S. R. K. Glanville, 'Working plan for shrine', *J. of Egyptian Archaeology* 16 (193 237–9; N. de G. Davies, 'An architect's pl from Thebes', *ibid.* 4 (1917), 194–9; and oth examples quoted by A. F. Shore, *op. cit.* (n.1
19 J. Wilkinson, *op. cit.* (n.17), i.382; so other expeditions are similarly drawn. Plans pharaonic tombs in the Valley of the Kings a preserved on an ostrakon in Cairo (G. Dares 'Ostraca', in *Catalogue général du Musée du Ca* I, Cairo, 1901, No. 25184) and a papyrus in t Museo Egizio, Turin (Howard Carter and H. Gardiner, 'The tomb of Ramesses IV and t Turin plan', *J. of Egyptian Archaeology* 4 (191 130–50).

20 P. Barguet, *Le Livre des Morts des anciens Egyptiens* (Paris, 1967), 28–30.
21 A. de Buck, *The Egyptian Coffin Texts* VII (Chicago, 1961), hierogl. texts 252–471, pls. 1–15; L. H. Lesko, *The ancient Egyptian Book of two Ways*: Univ. of Calif. Publ., Near Eastern Stud., 17 (Berkeley, Los Angeles and London, 1972); W. Bonacker, 'The Egyptian "Book of the two Ways"', *Imago Mundi* 7 (1950), 15–17.
22 G. Posener et al., *op. cit.* (n.14), p. 192; L. Casson, *Ancient Egypt* (Time-Life Netherlands, 1972), pp. 90–91.
23 O. Neugebauer and R. A. Parker, *Egyptian Astronomical Texts*, 3 vols. (Providence, R.I., 1960–69).
24 P. W. Pestman, *Greek and demotic Texts from the Zenon Archive* (Leiden, 1980), pp. 253–65, pl. XXIX. There are also, from Gebelein, fragments of a map of Pathyris, with inscriptions in Greek and demotic (Pap. Cairo demotic 31163).
25 C. C. Edgar, *Zenon Papyri in the University of Michigan Collection* (Ann Arbor, 1931), No. 84 and pl. VI.
26 P. Warren, 'The miniature fresco from the West House, Akrotiri, Thera, and its Aegean setting', *JHS* XCIX (1979), 115–29; S. Marinatos, *Excavations at Thera* VI (1974).
27 Catherine Delano Smith, 'The emergence of "maps" in European rock art', *Imago Mundi* 34 (1982), 9–25.
28 A. Brancati, *La biblioteca e i musei Oliveriani di Pesaro* (Pesaro, 1976), pp. 174–5.
29 Some reproductions of the Piacenza liver have the Etruscan writing either back to front or upside down.
30 *NH* ii.143, cf.142.
31 *Corpus agrimensorum Romanorum*, ed. C. O. Thulin (Stuttgart: Teubner, rev. edn. 1971), pp. 10–11; O. A. W. Dilke, 'Varro and the origins of centuriation', in *Atti del Congresso Internazionale di Studi Varroniani* (Rieti, 1976), pp. 353–8.
32 G. Körte, 'Die Bronzeleber von Piacenza', *Mitteilungen des kais. deutschen archäol. Inst., roem. Abt.* XX (1905), 348–77; C. O. Thulin, *Die etruskische Disciplin* II (Darmstadt, 1968); M. Pallottino, *Saggi di Antichità* II (Rome, 1979), 779–90.
33 Budge Collection, 89-4-26, 236.
34 *Nupt. Philol.* i. 45.
35 *Il.* xviii. 478–617; cf. Germaine Aujac, 'De quelques représentations de l'espace géographique dans l'Antiquité classique', *C.T.H.S. Bulletin de la Section de Géographie* 84 (1979), 27–38, esp. 27–8.
36 *Il.* xviii. 486–92.
37 *Ibid.* 607.

Chapter II

1 *Lives of the Philosophers* ii.2.
2 D. R. Dicks, *Early Greek Astronomy to Aristotle* (London, 1970), pp. 151–89; B. R. Goldstein and A. C. Bowen, 'A new view of early Greek astronomy', *Isis* 74 (1983), 330–40.
3 Ed. G. R. Mair (London, 1960).
4 *Poetae Latini Minores*, ed. E. Baehrens, i.1–28, 142 ff.; W. W. Ewbank, *The Poems of Cicero* (London, 1933, repr. New York and London, 1978); A. Traglia, ed., *Ciceronis Poetica Fragmenta* ii (Rome, 1952).
5 E. J. Dijksterhuis, *Archimedes* (Copenhagen, 1967).
6 Cic. *Rep.* i.22.
7 Miletus claimed to be a Cretan foundation, and Minoan-Mycenaean pottery found there lends weight to this claim.
8 Agathemerus i.1, in *Geographi Graeci Minores*, ed. C. Müller ii (Paris, 1882), 471; Diogenes Laertius ii.1–2; Herodotus ii.109.
9 See n.1.
10 Ps.-Plut. *Strom.* 2; Hippolytus, *Ref.* i.6.3; cf. G. S. Kirk and J. E. Raven, *The Presocratic Philosophers* (Cambridge, 1957), pp. 134–5.
11 v.49, 51.
12 *Fragmenta Historicorum Graecorum*, ed. C. and T. Müller (5 vols., Paris, 1878–85), i.1–31; G. Nenci, ed., *Hecataei Milesii fragmenta* (Florence, 1954).
13 Agathemerus (n.8 above).
14 Cicero, *De div.* i.1, says that Greek colonies were seldom founded without the advice and direction of the Delphic oracle.
15 Cf. p. 61 below.
16 Agathemerus (n.8), i.2.
17 *Phaedo* 109A–111C.
18 *Alcibiades* 17.4.
19 vi.1.1.
20 H. W. Catling, 'Archaeology in Greece, 1979–80', *Archaeological Reports for 1979–80* (Soc. Prom. Hell. Stud. and Brit. School at Athens), p. 12, col.2.
21 H. F. Mussche, *Thorikos: a Guide to the Excavations* (Brussels, 1974), pp. 50–51; the probable plan was discovered later.
22 i.1 (n.8).
23 E. Honigmann, *Die sieben Klimata und die* πόλεις ἐπίσημοι (Heidelberg, 1929).
24 *Introduction to Phaenomena* 16.5–6.
25 *Christian Topography* ii.80.

26 E.g. that of H. D. P. Lee (Loeb series); R. Böker, 'Winde: Windrosen', in *RE* VIII.A.2, cols. 2358–60.
27 C. Graux and A. Martin, 'Figures tirées d'un manuscrit des *Météorologiques* d'Aristote', *RPh* 24 (1900), 5–18, esp. 13–15, and pl.III.
28 Pliny, *NH* vi. 61.4, gives the name as *bematistae* (βῆμα = step).
29 C. F. C. Hawkes, *Pytheas: Europe and the Greek Explorers*, the eighth J. L. Myres Memorial Lecture (Oxford, 1975); R. Dion, 'Pythéas explorateur', *RPh* 92 (1966), 191–216.
30 But not Eratosthenes, who appreciated Pytheas' astronomical data.
31 In most Classical writers Mona is Anglesey. A. L. F. Rivet and Colin Smith, *The Place-names of Roman Britain* (London, 1979), pp. 41, 419, take it as always intended for Anglesey. But Caesar, *BG* v.13.2, is at best ambiguous, and the latitude given by Pytheas suits the Isle of Man better.
32 Strabo ii.4.2–3.
33 H. B. Gottschalk, 'Notes on the wills of the Peripatetic scholarchs', *Hermes* 100 (1972), 314 ff.
34 There is some indication that in trying to be up-to-date the librarians tended to throw out works thought to be superseded.
35 *Argonautica* iv. 259–93, esp. 279–81.
36 Agathemerus (n.8), ii.7.
37 R. M. Bentham, *The Fragments of Eratosthenes* (Ph.D. thesis, University of London, 1948); H. Berger, ed., *Die geographischen Fragmente des Eratosthenes* (1880, repr. Amsterdam, 1964).
38 F. Hultsch, *Griechische und römische Metrologie* (Berlin, 1882), index s.v. Stadion; *RE* s.v. Stadion: Metrologie.
39 'Calculating, as most do' (xvii.1.48). For Julian see F. Hultsch, ed., *Metrologicorum scriptorum reliquiae* (Leipzig: Teubner, 1864), i.201.9.
40 *History of Ancient Geography* (Cambridge, 1948), pp. 161–2 and notes.
41 Strabo i.4.1 ff.
42 O. Neugebauer and D. Pingree, *The Pañcasiddhāntikā of Varāhamiriha* ii (Copenhagen, 1971), 31, interpret a sixth-century AD Sanskrit text, based on earlier work, as giving the equivalent of 44° and 54° for the (longitudinal) distances from 'town of the Greeks', taken to be Alexandria, to Ujjayini (Ozene) and Benares. The correct distances are $45^\circ 50'$ and $53^\circ 7'$; such acccuracy would seem to point to lunar eclipse observations.
43 Strabo ii.3.2.
44 Suet. *Gram.* 2.
45 Strabo ii.5.10 says that to represent the eart one should make a globe like that of Crates, an that to show the oikumene clearly a globe of a least 10 ft in diameter is needed.
46 *Od.* i.22–4.
47 F. M. Snowdon, *Blacks in Antiquit* (Cambridge, Mass., 1970), pp. 101–5.
48 Macr. *Comm.* v.31–6.
49 D. R. Dicks, *The Geographical Fragments o Hipparchus* (London, 1960).
50 Frs. 35, 36, 39 Dicks.
51 *NH* ii.108; cf. Fr. 38 Dicks, with note on p 153.
52 *Aren.* i, 8.
53 Strabo ii.5. 7.
54 D. J. Campbell, ed., C. Plini Secund *Naturalis Historiae liber secundus* (Aberdeer Univ. Studies, No. 118), 1936, p. 91.
55 Frs. 19–22 Dicks, with commentary pp 130–35.
56 *Op. cit.* (n.49), pp. 130–37.
57 Fr. 61 Dicks, with commentary pp. 185–91
58 For Eratosthenes' measurement see p. 32 Posidonius, based on Rhodes, will not have been so influenced by Egyptian measurements

Chapter III

1 Plin. *NH* xxxv. 11.
2 A. Riese, ed., *Geographi Latini minore* (Heilbronn, 1878, repr. Hildesheim, 1964), alsc Dicuil and the Hereford World Map.
3 The figures relating to years of completion are expressed in terms of consulships.
4 A. Klotz, 'Die geographischen commentarii des Agrippa und ihre Überreste', *Klio* 25 (1931), 35ff., 386ff.
5 Mart. ii.14.3, 5,15; iii.20.12; vii.32.11; xi.1. 11.
6 It is clear that in addition to Augustus' librarian there were many pseudo-Hygini; see p. 100 below.
7 D.(S.D.F.) Detlefsen, *Ursprung, Einrichtung und Bedeutung der Erdkarte Agrippas* (Berlin, 1906).
8 A. Riese, *op. cit.* (n.2).
9 *NH* iv.50.
10 Y. Janvier, *La géographie d'Orose* (Paris, 1982), p. 266.
11 *Liber de mensura orbis terrae*, ed. J. J. Tierney: Scriptores Latini Hiberniae (Dublin, 1967).
12 In *NH* v.67, as pointed out by Klotz, *op. cit.* (n.4), 44, the length of Syria is basically north–south; and Detlefsen seems to have been

wrong in thinking that Pliny by mistake inverted its length and breadth.
13 MS variants on 80 are 75 and 70.
14 vii.3.19.
15 D.(S.D.F.) Detlefsen, op. cit. (n.7).
16 The Latin is *usquequaque transiit* (MSS *transit*).
17 Pline l'Ancien, *Histoire Naturelle*, Livre v, 1–46, ed. J. Desanges (Paris: Budé collection, 1980).
18 W. Aly, 'Die Entdeckung des Westens', *Hermes* 62 (1927), 299–341, esp. 338.
19 v.41.
20 Strabo ii.1.39.
21 The MSS have *Perico* for *Serico*, which can be shown from Dicuil to be the correct reading.
22 An alternative reading, supported by Martianus Capella, is 1722.
23 *Op. cit.* (n.4).
24 Klotz does not imply that Eratosthenes took 8 stades to the mile, merely that Romans would automatically apply this equivalent.
25 *Geographi Latini minores*, p. IX.
26 The numbers add up to 4708, which has been proposed as an amendment; see Klotz, *op. cit.* (n.4).
27 Other suggestions, apart from Julius Caesar (p.40 above), are Varro (G. Öhmichen, *Plinianische Studien* (Erlangen, 1880), pp. 58ff.) and an anonymous Roman chorography postulated by A. Riese, *op. cit.* (n.2), pp. VIII–XVII.
28 iv.3.33–40.
29 British Library, Harl. 2767; see F. Granger, ed., Vitruvius vol. 1 (Loeb), pp. xvi ff.
30 Vitr. i.6.12 *visum est mihi in extremo volumine formas (formam H), sive ut Graeci schemata dicunt, duo explicare, unum ita deformatum ut appareat unde certi ventorum spiritus oriantur, alterum quemadmodum ab impetu eorum aversis derectionibus vicorum et platearum evitentur nocentes flatus.* In Granger, *op. cit.* (n.29), pl. A, the upper figure is reproduced from H (Harl. 2767), the lower one reconstructed from the text.
31 viii.2.6.
32 *XII Panegyrici Latini*, ed. R. A. B. Mynors (Oxford, 1964), IX (IV).

Chapter IV

1 *The Authoress of the Odyssey* (London, 1897).
2 W. Dörpfeld was convinced that it was not Ithaki but Lefkas. He excavated Mycenaean houses at Nidri on that island, whose inhabitants have erected a statue to him. But apart from the linguistic difficulties, such an identification rests on a probable failure to appreciate Homeric turns of phrase.
3 See p. 20 above and p.145 below.
4 Hom. *Il.* xviii. 607–8.
5 i.1 (*Geographi Graeci minores*), ed. C. Müller, ii.471).
6 iv. 36. An alternative theory has it that Hyperboreans are really 'men who pass on', referring to the legend that every year they handed on, through a succession of tribes, ears of wheat as an offering to Apollo on the island of Delos (Herodotus iv.33; Call. *Del.*283–4).
7 Agathemerus i.1(n.5).
8 Herodotus iv. 37–45.
9 Europa was famous in mythology; Asia is said to have been one of the Oceanides; Libya, however, does not seem to occur as the name of a woman. Waldseemüller or his associate Martin Ringmann was evidently recalling this passage of Herodotus when he suggested America as the name of the New World (p. 165).
10 iv.45.
11 iii.115.
12 iv.42–3; cf. p. 24.
13 ii.10–11.
14 Philostr. *Imag.* i.9.
15 Tides in the Mediterranean are minimal except near the island of Djerba, so that oceanic tides were a keen object of speculation.
16 ii.19ff.
17 Agathemerus i.2 (n.5).
18 *NH* v.9.
19 Strabo ii. 4. 1–3, cf. F. W. Walbank, *A Commentary on Polybius*, vol. 3 (Oxford, 1979), pp. 630–32.
20 Strabo viii. 8.5. W. R. Paton, in his translation of Polybius xxxiv.12.12, wrongly gives 1000 stades.
21 ii. 2. 1–3.
22 According to Posidonius, Parmenides was the originator of this theory; according to Plutarch, *De placitis philos.* 2.12 and 3.14, it was Pythagoras.
23 Strabo ii. 3.4.
24 Transl. H. L. Jones, 10 vols. (Loeb series, Cambridge, Mass., and London, 1917–33); Germaine Aujac, *Strabon et la science de son temps* (Paris, 1966).
25 This Tyrannio, a freedman of Cicero's, was a pupil of Tyrannio of Amisos, a grammarian.
26 See n.23.
27 i.2.9.
28 i.2.39.
29 *Il.* xiv.200.
30 *Od.* v.393.
31 Strabo i.2.22 ff.

32 *Od.* i.23.
33 ii.3.8.
34 The novelist Heliodorus, for example, writing his *Aethiopica* under the late Roman Empire, called his wise men of Ethiopia gymnosophists, 'naked fakirs', a term more generally used of Indians.
35 ii.5.4.
36 ii.3.1.
37 This was only an area where cinnamon was traded: it actually came from South or South-east Asia.
38 ii.1.13–17, cf. i.4.2, ii.2.2. Like many other ancient geographers, Strabo thought of Ireland as much further north than it is.
39 i.4.4.
40 ii.5.5–6. Strabo's view of the extreme limits of the habitable earth was derived from that of Eratosthenes (see fig.4).
41 The term 'parallels' is used, with reference to Graeco-Roman cartography, of lines connecting places or areas regarded as being on the same latitude. By 'circles' Strabo means arcs of circles.
42 ii.5.10.
43 Pliny, *NH* v.51–2.
44 *Ibid.*vi.203–5.
45 Ed. G. Ranstrand (Göteborg, 1971); id., *Textkritische Beiträge zu Pomponius Mela* (Göteborg, 1971); ed. P. G. Parroni (Rome, 1984).
46 *Geo.* i.233–9. Eratosthenes' original is preserved for the portion corresponding to 235–9, cf. Conington's edn. of Virgil *ad loc.* The idea of five such zones goes back to Aristotle.
47 iii.31.
48 Müllenhoff conjectured *Scadinavia*, a form found elsewhere.
49 iii.54.
50 Nicolaus Sallmann, 'De Pomponio Mela et Plinio maiore in Africa describenda discrepantibus', *Africa et Roma, Acta omnium gentium ac nationum Conventus Latinis litteris linguaeque fovendis*, ed. G. Farenga Ussani (Rome, 1979), pp. 164–73.
51 The usual modern view is that Tarshish was a district rather than a town.
52 Transl. H. Rackham *et al.* (Loeb series, 10 vols. (London and Cambridge, Mass., 1938–62; Budé edn., with French translation, 24 vols. to date (Paris, from 1950). Most editions of Pliny's *Natural History* are numbered in chapters as well as sections, but modern usage is to omit chapters in references.
53 K. G. Sallmann, *Die Geographie des älteren Plinius in ihrem Verhältnis zu Varro* (Berlin and New York, 1971), pp. 176–7 quotes other passages too in which Pliny praises Gree scientific achievements.
54 His Latin world is *circuli* (Greek *parelleloi* These differ from *klimata* in that they can appl to any circles of latitude or longitude.
55 Strabo, in the Alexandrian tradition, star his southernmost zone at Meroë (Sudan).
56 For the structure of Pliny's work see K. G Sallmann, *op. cit.* (n.53), pp. 191ff.
57 *NH* iii.5.
58 *Ibid.* 39.
59 *NH* iv.102–3.
60 *NH* vi.200. He also quotes Hanno as source, and refers in ii.69 to the account of hi circumnavigation.
61 *NH* v.11–15.
62 We learn from Pliny, *NH* v.12, that *equites* who later pillaged the province for luxuriou timbers and dyes, boasted of this penetration He is scornful of such *equites*, who in his day even get into the senate; so that his phrase *i gloria fuit* does not mean that he praises them fo their explorations, but that he doubts thei veracity when they made this boast.
63 *NH* v.15.
64 R. K. Sherk, 'Roman geographical exploration and military maps', *Aufstieg und Niedergang der römischen Welt*, II, 1, ed. H. Temporin (Berlin and New York, 1974), 534–62.
65 *NH* v.66ff.
66 *NH* vi.36–8.
67 *Ibid.* 60.
68 *Ibid.* 81–3.
69 *Ibid.* 101–6.
70 *Ibid.* 117–41.
71 *Ibid.* 141.
72 Not his son, as Pliny, *NH* vi.160, gives.
73 *NH* vi.162.
74 *Ibid.* 178–205.

Chapter V

1 O. Neugebauer, 'Über eine Methode zu Distanzbestimmung Alexandria-Rom be Heron', *Kgl. Danske Vidensk. Selsk., Hist.-filol Medd.* 26.2 (1938), 26.7 (1939).
2 Marinus is said by Ptolemy, *Geog.* i.6.1. to be 'the latest' of cartographers, hence not long before his time: cf. *RE* XIV.ii s.v. Marinos 2, *RE* Suppl. XII s.v. Marinos.
3 *Geog.* i.6.1: the Greek is διόρθωσις.
4 *Geog.* i.17.1; i.18.3.
5 *Geog.* i.5.
6 *Geog.* i.7.
7 Ptol. *Geog.* i.7.6.
8 *Ibid.* i.8.1.

9 J. Oliver Thomson, *History of Ancient Geography* (Cambridge, 1948), p. 266.
10 Ptol. *Geog.* i.11.
11 *Ibid.* i.14.
12 *Ibid.* i.15.
13 Stane Street (Margary 15): I. D. Margary, *Roman Roads in Britain* (London, rev. edn. 1967), pp. 58ff.; A. L. F. Rivet, 'Viae aviariae?' *Antiquity* LVI (1982), 206–7.
14 Ed. J. L. Heiberg, 2 vols. (Leipzig: Teubner, 1898–1903); Eng. trans. by C. J. Toomer, *Ptolemy's Almagest* (London, 1984).
15 *Synt. Math.* i.4.
16 E. Honigmann, *Die sieben Klimata und die* πόλεις ἐπίσημοι (Heidelberg, 1929).
17 *Synt. Math.* ii.13.
18 *Geog.* i.1. Edn of whole work, C. F. A. Nobbe (Leipzig, 1843–5); of Books I–VI, C. Müller (Paris, 1883–1901); Eng. trans., not always accurate, by E. L. Stevenson (New York, 1932).
19 Review of Halma's edn. of Ptolemy's *Geography*, in *Journal des Savants* 1830, p. 742.
20 The word πόλις (*polis*) tends in the *Geography* to mean any sort of settlement.
21 *Ann.* iv.73, cf. Ptol. *Geog.*ii.11.27.
22 *Geog.* i.21.
23 *Geog.* ii.1.2.
24 *Geog.* i.24.9–20.
25 *Geog.* i.24.9.
26 *Geog.* vii.6–7; cf. O. Neugebauer, 'Ptolemy's Geography, Book VII, Chapters 6 and 7', *Isis* 50 (1959), 22–9.
27 *Geog.* viii.3.1 ff.
28 In favour of the view that Ptolemy's *Geography* came out with maps: P. Dinse, *Zeitschrift der Gesellschaft für Erdkunde* 1913, 754ff. and *Zentralblatt für Bibliothekswesen* XXX (1913), 389ff.; O. Cuntz, ed., *Die Geographie des Ptolemaeus . . .* (Berlin, 1923), pp. 24ff.; J. Fischer, ed., *Claudii Ptolemaei Geographiae Codex Urbinas Graecus 82* (Leiden, Leipzig and Turin, 1932), I.1.108, 130ff.; P. Schnabel, *Text und Karten des Ptolemaeus* (Leipzig, 1938), pp. 95ff.; L. A. Brown, *The Story of Maps* (1949, repr. New York, 1977), p.68; E. Polaschek, 'Ptolemy's *Geography* in a new light', *Imago Mundi* 14 (1959), 17–37, who thought that Ptolemy drew regional maps, not a world map, and for his first edition only; R. Baladié, *Le Peloponnèse de Strabon* (Paris, 1980), p.18 and n.3. Against: H. Berger, *Geschichte der wissenschaftlichen Erdkunde der Griechen* (Leipzig, 2nd edn. 1903), pp. 640f.; K. Kretschmer, *Zeitschrift der Gesellschaft für Erdkunde* 1913, 767ff.; 1914, 783; id., *Petermanns Geographische Mitteilungen* 60 (1914), 142f.; L. O. T. Tudeer, *JHS* XXXVII (1917), 66; F. Gisinger in Pauly-Wissowa, *RE* Suppl. IV (1924), col.666; L. Bagrow, 'The origins of Ptolemy's Geographia', *Geografiska Annaler* 1945, 318–87; id., *Meister der Kartographie* (Berlin, 1963), pp. 39–42. Uncommitted: W. Kubitschek in Pauly-Wissowa, *RE* X (1919), cols.2088ff.; E. Honigmann, *Klio* XX (1925), 204, but in Pauly-Wissowa, *RE* XIV (1920), col.1771 against; J. N. Wilford, *The Mapmakers* (London, 1981), p. 26.
29 Cf. also Cassiodorus (p. 155 below).
30 *RE* Suppl. X, col. 737.
31 L. Bagrow, 'The origin of Ptolemy's Geographia', *Geografiska Annaler* 1945, 318–87.
32 *Claudii Ptolemaei Geographiae Codex Urbinas Graecus 82* (Leiden, Leipzig and Turin, 1932).
33 For a convenient summary see A. L. F. Rivet and Colin Smith, *The Place-names of Roman Britain* (London, 1979), pp. 130–31.
34 Thus Seneca, *N.Q.* i, *prol.* 13, says that the journey from Spain to India is only a very few days if the ship gets a perfect wind.
35 J. Oliver Thomson, *op. cit.* (n.9), pp. 315–19.
36 F. Hirth, *China and the Roman Orient* (Leipzig, 1885, repr. Shanghai and Hong Kong, 1939), p.47; M. Cary and E. H. Warmington, *The ancient Explorers* (London, 1929), pp. 82–3; *RE* s.v. Cattigara. For contacts with China see J. Ferguson, 'China and Rome', *ANRW* II.9.581–603; M. G. Raschke, 'New studies in Roman commerce with the East', *ibid.* 604–1378.
37 *RE* Suppl. X., cols.711–34.
38 The actual figures are 4°51′ W. (= 13°01′ E. of La Palma, Canaries); 54°38′ N.
39 A. L. F. Rivet, 'Ptolemy's Geography and the Flavian invasion of Scotland', in *Studien zu den Militärgrenzen Roms* II (Vorträge des 10. internationalen Limeskongresses in der Germania Inferior), Cologne, 1977, pp. 45–54, with bibliography.
40 v.21.4.
41 Ed. C. Müller, 2 vols. and Tabulae (Paris, 1883–1901).
42 Rivet and Smith, *op. cit.* (n.33), pp. 103–47.
43 *Map of Roman Britain*, 4th edn. (Southampton, 1978).
44 The modern (1902) town of Morecambe was given a name based on the Ptolemaic estuary: there is no survival of name from antiquity, the earlier name having been Poulton-le-Sands.
45 Cf. Rivet and Smith, *op. cit.* (n.33), p. 119, Table 1.
46 *RIB* 70.

47 I. A. Richmond and O. G. S. Crawford, 'The British section of the *Ravenna Cosmography*', *Archaeologia* XCIII (1949), 1–50; cf. Rivet and Smith, *op. cit.* (n.32), pp. 185–215.

48 *An Assessment of the Value of some Literary Sources for Roman Britain*, unpublished M. Phil. thesis, Southampton, 1974; cf. Rivet and Smith, *op. cit.* (n.33), pp. 123–4.

49 *Ibid.* p. 483.

50 Rivet and Smith's suggestion is Charterhouse, at 5 km perhaps too distant from the old course of the Axe. All that is known from the Roman period at the mouth of the Axe is a shrine.

51 *RE* Suppl. X, cols.711ff.

52 Ptol. *Geog.* vii.2.1, 7, 20; cf. vii.3.3.

53 D. E. Ibarra Grasso, *La representación de América en mapas romanos de tiempos de Cristo* (Buenos Aires, 1970).

Chapter VI

1 See photograph in O. A. W. Dilke, *The Roman Land Surveyors: an Introduction to the Agrimensores* (Newton Abbot, 1971), p. 208. This book has been translated into Italian, slightly revised and with more illustrations, as *Gli Agrimensori di Roma antica* (Bologna, 1979). A simple kit, O. A. W. Dilke, *Surveying the Roman Way* (University of Leeds, 1980), gives instructions for using the groma and other Roman surveying instruments.

2 F. Castagnoli, *Ippodamo di Mileto e l'urbanistica a pianta ortogonale* (Rome, 1956), and English translation, *Orthogonal Town Planning in Antiquity*, tr. V. Caliandro (Cambridge, Mass., 1971).

3 Hector Williams, 'Stymphalos: a planned city of classical Arcadia', summary in XIIe Congrès International d'Archéologie Classique, Athènes, 4–10 septembre, 1983, p. 176.

4 G. Schmiedt and R. Chevallier, 'Caulonia e Metaponto', *L'Universo* 39 (1959), 349–70.

5 γεωμέτρης Lucian, *Icar.* 28; earlier in Doric form γαμέτρας (nom. sing.) in the Heraclea Tablets.

6 F. Castagnoli, *op. cit.* (n.2).

7 There are also other names, of which in the late Empire the commonest was *gromatici*, 'groma men'.

8 The metal parts of the only instrument which is certainly a groma were found at Pompeii and are now in a technical room, opened on request, in the Naples Archaeological Museum: M. Della Corte, 'Groma', *Monumenti Antichi* 28 (1922), 5–100.

9 C. V. Walthew, 'Property-boundaries and the sizes of building-plots in Roman towns', *Britannia* 9 (1978), 335–50; id., 'Possible standard units of measurement in Roman military planning', *Britannia* 12 (1981), 15–35; P. Crummy, 'The origins of some major Romano-British towns', *Britannia* 13 (1982), 125–34.

10 B. M. *Catalogue of Bronzes* 2677.

11 F. Castagnoli, *Le ricerche sui resti della centuriazione* (Rome, 1958); O. A. W. Dilke, *op. cit.* (n.1), pp. 133–58; id., 'Archaeological and epigraphic evidence of Roman land survey', in *Aufstieg und Niedergang der römischen Welt* ii.1 (1974), 564–92; G. Chouquet and F. Favory, *Contribution à la recherche des cadastres antiques*: Annales littéraires de l'Université de Besançon, 236 (Paris, 1980).

12 V. M. Rosselló Verger et al., *Estudios sobre centuriaciones romanas* (Cantoblanco, 1974).

13 O. A. W. Dilke, 'De agrorum divisione in provinciis septentrionalibus ad occidentem vergentibus', *Acta Treverica 1981* (Leichling, 1984), 47–51.

14 Published by the Institut Géographique National (Paris, 1954).

15 James Nelson Carder, *Art historical Problems of a Roman Land Surveying Manuscript: the Codex Arcerianus A, Wolfenbüttel* (New York and London, 1978).

16 O. A. W. Dilke, 'Illustrations from Roman surveyors' manuals', *Imago Mundi* XXI (1967), 9–29.

17 F. Castagnoli, *op. cit.* (n.11), p. 10, n.2; O. A. W. Dilke, 'Maps in the treatises of Roman land surveyors', *Geographical J.* 127 (1961), 417–26.

18 Pliny, *NH* iii.59.

19 *Possessores* were occupiers rather than owners, since *possessio* did not imply legal ownership.

20 *Constitutio limitum* (*Corpus agrimensorum* ed. C. Thulin, p. 143).

21 A. Schulten, 'Römische Flurkarten', *Hermes* 33 (1898), 534–68.

22 G. Grosjean, 'La limitation romaine autour d'Avenches', *Le Globe* 95 (1956), 57–74.

23 For the town plan of Aosta, substantially preserved in the modern layout, see J. B. Ward Perkins, *Cities of Ancient Greece and Italy: Planning in Classical Antiquity* (London, 1974), pp. 28, 120, figs 52–3.

24 Virg. *Ecl.* ix.27–9.

25 *Constitutio limitum* (*Corpus agrimensorum*, ed. C. Thulin, p. 162).

26 *De controversiis* (*ibid.*, p. 10, ll.5ff.).
27 *De agrorum qualitate* (ibid., p. 2. ll.8ff.).
28 *Constitutio limitum* (ibid., pp. 159–60).
29 *Ibid.* p. 160, ll.8ff.
30 *Ibid.* p. 165, ll.10ff.
31 Segustero is Sisteron in the Alpes de Haute Provence; the closest corruption, to supply a name in Italy, is Segusio (Susa).
32 *Constitutio limitum* (*Corpus agrimensorum*, ed. C. Thulin, p. 166, ll.3–6, 10–15; p. 167, ll.5–12).
33 F. Blume et al., ed., *Die Schriften der römischen Feldmesser* (Berlin, 1848–52), repr. n.d.
34 O. A. W. Dilke, *op. cit.* (n.16), 17–18, with reconstructed diagram by C. Koeman.
35 Frontinus, *De aquae ductu urbis Romae*, ed. C. Kunderewicz (Leipzig: Teubner, 1973).

Chapter VII

1 J. Mellaart, *Çatal Hüyük: a Neolithic Town in Anatolia* (London, 1967), figs. 59–60 and for actual landscape pl.I.
2 W. Blumer, 'The oldest known plan of an inhabited site dating from the Bronze Age', *Imago Mundi* 18 (1964), 9–11; Catherine Delano Smith, 'The emergence of "maps" in European rock art', *ibid.* 34 (1982), 9–25.
3 J. J. Coulton, *Greek Architects at Work* (London, 1977), pp. 52ff. investigates the likelihood of Greek architects having used plans, and thinks they were little used before the Hellenistic period.
4 Vitr. *De architectura* ii. praef.1. Dinocrates proposed the founding of a major new settlement on Mt Athos, and held a model of the walled city in his left hand.
5 *De architectura* i.2.2.
6 P. Camus, *Le pas des légions* (Paris, 1974), front cover, and for plan of fort p. 62.
7 Measurements taken by their orders in AD 73 are quoted by Pliny, *NH* iii.66–7; cf. G. Carettoni et al., *La pianta marmorea di Roma antica*, 2 vols (Rome, 1960), p. 218 and notes.
8 O. A. W. Dilke, 'Ground survey and measurements in Roman towns', *British Council of Archaeology: Research Reports* (forthcoming).
9 H. Stuart Jones, *A Catalogue of the ancient Sculptures preserved in . . . Rome: the Sculptures of the Museo Capitolino*, 2 vols. (Oxford, 1912), pl.15 (II.8).
10 G. Carettoni et al., *op. cit.* (n.8).
11 The wording is SEVERI ET ANTONINI AVGG NN. The use of NN for *nomina* is as in the Book of Common Prayer, where 'N. or M.' stands for 'N. or NN'.
12 See E. Rodriguez Almeida, *Forma Urbis marmorea: aggiornamento generale 1980*, 2 vols. (Rome, 1981).
13 Vaticanus latinus 3439; cf. Carettoni et al., *op. cit.* (n.7); A. M. Colini and L. Cozza, *Ludus Magnus* (Rome, 1962).
14 F. Castagnoli, 'L'orientamento nella cartografia greca e romana', *Rendiconti della Pontificia Accademia di Archeologia XLVIII* (1975–6), 59–69.
15 *Hist.* iii.82.
16 Suet. *Aug.* 28.
17 An indication that land surveyors were conscious of scale is given by a title to part of the *Casae litterarum: incip. et de casis litterarum montium: in ped. v̄ fac pede uno*, interpreted as 1:5,000: J. N. Carder, *Art historical Problems of a Roman Land Surveying Manuscript* (New York and London, 1978), p. 160.
18 Rodriguez Almeida, *op. cit.* (n.12), i.115–18.
19 The fragments depicting the *vicus patricius* have been very probably fixed: Rodriguez Almeida, *op. cit.* (n.12), i. 86–7.
20 Among researchers on the Forma Urbis in this century have been Hulsen, Lundström, Gatti, Carettoni and his associates, and Rodriguez Almeida.
21 It was earlier erroneously identified as a temple.
22 V. Lundström, *Undersökningar i Roms Topografi* (Göteborg, 1929).
23 G. Gatti, 'Topografici dell' Iseo Campense', *Rendiconti della Pontificia Accademia Romana di Archeologia* XX (1943–4), 117ff.
24 Recent excavations in the area of the Crypta Balbi have revealed serious discrepancies between the buildings and streets excavated and the Forma Urbis. This, however, constitutes at present the only area where such substantial defects can be proved.
25 Both the fragments mentioned below are illustrated in P. D. A. Harvey, *The History of Topographical Maps* (London, 1980), p. 130. The second is in the Ostia museum, Room I, Inv. 191: fragment of plan with shops, from tomb of Julia Procula, with measurements in Roman feet: CLVII, CC, XIX.
26 Einhard, *Vita Karoli*, 33.
27 Carettoni et al., *op. cit.* (n.7), pp. 207–8, No. 3 and Tav. agg. Q, fig. 51; *CIL VI.* 29847.
28 *Ibid.* p. 209, No. 7.
29 *Les documents cadastraux de la colonie romaine d'Orange* (*Gallia* Suppl. 16), Paris, 1962; O. A. W. Dilke, *The Roman Land Surveyors* (Newton Abbot, 1971), pp. 159–77.
30 J. H. Oliver, 'North, South, East, West at

Arausio and elsewhere', in *Mélanges d'archéologie et d'histoire offerts à A. Piganiol* II. 1075–9 (Paris, 1966).

31 O. A. W. Dilke, 'The Arausio cadasters', *Vestigia* 17 (1973), 455–7 (Akten des VI. Kongresses für griech. u. röm. Epigraphik).

32 *CIL* VI.1261.

33 *CIL* VIII.2728.

34 The name 'Boscovich' refers to R. G. Boscovich, who helped the original owner P. M. Paciaudi to interpret the anemoscope. See I. Zicàri, 'L'anemoscopio Boscovich del Museo Oliveriano di Pesaro', *Studia Oliveriana* II (Pesaro, 1954), 69–75; R. Böker, in *RE* VII.A.2, cols. 2358–60, s.v. Winde: Windrosen; A. Brancati, *La Biblioteca e i Musei Oliveriani di Pesaro* (Pesaro, 1976), 102, 210, Tav. XLII.

Chapter VIII

1 Vegetius, *Epitoma rei militaris* iii.6. Hence the title of Annalina and Mario Levi's book, *Itineraria picta* (Rome, 1967). There is, however, some doubt about the meaning. Vegetius' words are: . . . *usque eo ut sollertiores duces itineraria provinciarum in quibus necessitas gerebatur [geritur vulgo] non tantum adnotata sed etiam picta habuisse firmentur, ut non solum consilio mentis verum aspectu oculorum viam profecturus eligeret.* This could refer to paintings of landscape, as suggested in a seminar by P. Arnaud.

2 A section of the forthcoming *History of Cartography* (Chicago), vol. 1 will be devoted to this topic.

3 Ekkehard Weber, *Tabula Peutingeriana: Codex Vindobonensis 324* (Graz, 2 vols., 1976); Konrad Miller, *Die Peutingersche Tafel* (repr. Stuttgart, 1962); Annalina and Mario Levi, *Itineraria picta: contributo allo studio della Tabula Peutingeriana* (Rome, 1967).

4 An interesting precedent, some 800 years earlier, for the use of glass partly as a preservative and partly to depict seas, comes from the Latin poem by Baudri mentioned below (p. 153).

5 *Itineraria Romana* (Stuttgart, 1916), Preface and elsewhere.

6 H. Gross, *Zur Entstehungs-Geschichte der Tabula Peutingeriana*, diss. Friedrich-Wilhelms Universität, Berlin, 1913.

7 Theodor Mommsen, *Gesammelte Schriften* V. 305, went so far as to think that Agrippa's map contained roads; but this is quite unproven.

8 C. H. Roberts, 'The Codex', *Proceedings of the British Academy* 40 (1954), 169–204; O. A. W. Dilke, *Roman Books and their Impact* (Leeds, 1977), pp. 24–5. The parchment world map which Mettius Pompusianus carried round with him in Domitian's reign was probably not typical.

9 The greatest length surviving is Egyptian, Papyrus Harris I, which is 43.6 m long.

10 There is now clear evidence that in the late Empire a small settlement was re-established in parts of Pompeii.

11 A. de Franciscis, 'La villa Romana di Oplontis', in B. Andreae and H. Kyrieleis, ed., *Neue Forschungen in Pompeii* (Recklinghausen, 1975); C. Malandrino, *Oplontis* (Naples, 1977).

12 W. Kubitschek, 'Eine römische Strassenkarte', *Jahresheften des Österreichischen Archaeologischen Instituts* 5 (1902), 20–96.

13 The argument is best put by A. and M. Levi, *op. cit.* (n. 3), pp. 97ff. For the *cursus publicus* see their n.104, and E. Badian, 'Postal Services', in *The Oxford Classical Dictionary*, 2nd edn. (Oxford, 1970).

14 *Corpus Inscriptionum Latinarum* VI.1774 = Dessau 5906.

15 A. and M. Levi, *op. cit.* (n.3), pp. 65ff.

16 Some of the villas shown on north African mosaics have façades somewhat similar to these. See A. and M. Levi, *op. cit.* (n.3), pp. 88f.; Katherine M. D. Dunbabin, *The Roman Mosaics of North Africa* (Oxford, 1978), pl. XLIII. 109; XLIV. 111–12; XLV. 113.

17 K. Miller, *Itineraria Romana* (Stuttgart, 1916); A. and M. Levi, *op. cit.* (n.3), pp. 17ff., 151ff. with other references.

18 G. Calza, 'La figurazione di Roma nell'arte antica', *Dedalo* 7 (1927), 663–88.

19 The alternative explanation, a lighthouse, seems unlikely if one compares its appearance with that of the three conventional signs on the Peutinger Table which undoubtedly represent lighthouses.

20 A. L. F. Rivet and Colin Smith, *The Place-names of Roman Britain* (London, 1979), pp. 149–50 and pl. I.

21 N. G. L. Hammond, *A History of Macedonia* i (Oxford, 1972), 128–34; cf. C. F. Edson, 'Strepsa', *Classical Philology* 50 (1955), 169–90.

22 E. Oberhummer, *Aus Nordgriechenland und Arkadien* (Leipzig, 1900).

23 Livy xliv.40.8; Plutarch, *Aem.* 16.5.

24 R. K. Sherk, 'Roman geographical exploration and military maps', in H. Temporini, ed., *ANRW*, II.1 (Berlin, 1974), 534–62 is wrong (p.561) in saying that Roman surveying maps are in 'the same bizarre proportions' as the Peutinger Table; their proportions are not in the least comparable.

25 F. Cumont, 'Fragment de bouclier portant une liste d'étapes', *Syria* 6 (1925), 1–15, pl. 1; F. Cumont, *Fouilles de Doura Europos, 1922–1923* (Paris, 1926), pp. 323–37, pls. CIX, CX.
26 R. Uhden, 'Bemerkungen zu dem römischen Kartenfragment von Dura Europos', *Hermes* 67 (1932), 117–25.
27 *RE* s.v. Itineraria.
28 R. Chevallier, *Les voies romaines* (Paris, 1972), pp. 46–9; Engl. transl., *Roman Roads* (London, 1976), pp. 47–50.
29 Both the *fines* wording and the Cottian Alps appear on only one of the four goblets; if it is the latest, it could date from not long after the work by Cottius.
30 J. Heurgon, 'La date des gobelets de Vicarello', *Revue des Etudes Anciennes* 54 (1952), 39–50.
31 *Geographi Graeci minores*, ed. C. Müller, i. 244ff. and introd. pp.lxxx ff.
32 J. D. Cowen and I. A. Richmond, 'The Rudge Cup', *Archaeologia Aeliana*, 4th ser. 12 (1935), 310–42.
33 J. Heurgon, 'The Amiens *patera*', *JRS* 41 (1951), 22–4.
34 The chief manuscripts are: D = Paris.Lat.7230A (10th century); L = Vienna, Nationalbibliothek 181 (8th century); P = Escorial II R 18 (8th century).
35 D. von Berchem, 'L'annone militaire dans l'Empire romain au IIIe siècle', *Mémoires de la Société Nationale des Antiquaires de France* 24 (1936), 117–201, esp. 166–81.
36 A. L. F. Rivet, 'The British section of the Antonine Itinerary', *Britannia* 1 (1970), 34–68; A. L. F. Rivet and Colin Smith, *The Place-names of Roman Britain* (London, 1979), pp. 150–80.
37 The Malton fort is well preserved, and most scholars say or imply that it is Derventio: see Rivet and Smith, *op. cit.* (n.36), pp. 333–4; J. F. Robinson, *The Archaeology of Malton and Norton* (Leeds, 1978), p. 8 with n.54. However, as pointed out by R. Selkirk, *A Dramatic new View of Roman History: the Piercebridge Formula* (Cambridge, 1983), pp. 67–9, Stamford Bridge is the only logical equivalent if the figure vii is correct. It is always possible that a Roman fort may be discovered there. If Derventio is Stamford Bridge, Delgovicia will be Malton.
38 K. Miller, *Itineraria Romana* (Stuttgart, 1916); O. Cuntz, *Itineraria Romana* I (Leipzig: Teubner, 1929); Ordnance Survey, *Map of Roman Britain*, 4th edn. (Southampton, 1979).
39 W. Rodwell, 'Milestones, civic territories and the Antonine Itinerary', *Britannia* 6 (1975), 76–101.
40 P. Geyer, ed., *Itinera Hierosolymitana saec.IIII–VIII* (Corpus Scriptorum Ecclesiasticorum Romanorum 39, 1898).
41 *Satires* i.5.
42 L. Casson, *Travel in the Ancient World* (London, 1974), p. 190.

Chapter IX

1 *RE* s.v. Periplus; *Geographi Graeci minores*, ed. C. Müller (3 vols., Paris, 1882); A. E. Nordenskiöld, *Periplus*, transl. Francis A. Bather (Stockholm, 1897).
2 The Romans were content to translate from Punic into Latin such works as the *Agriculture* of Mago.
3 V. Bérard, *Les Phéniciens et l'Odyssée*, 2 vols. (new edn., Paris, 1927); A. Shawan, *Homeric Essays* (Oxford, 1935); H. H. and A. Wolf, *Der Weg des Odysseus* (Tübingen, 1968); W. B. Stanford and J. V. Luce, *The Quest for Ulysses* (London, 1974).
4 *Od.* vii.253 and elsewhere.
5 Nights in the country of the Laestrygones were so short that men taking out animals to pasture met men bringing animals back (Hom. *Od.* x.82ff.). But the Cimmerians seem to have originated no further north than south Russia.
6 Rufus Festus Avienus (more correctly Avienius), *Ora Maritima*, ed. J. P. Murphy (Chicago, 1977), p. 54, note on 154–7.
7 C. F. C. Hawkes, *Pytheas: Europe and the Greek Explorers*: the eighth J. L. Myres memorial lecture (Oxford, 1975).
8 *Geographi Graeci minores* (n.1), i. xvii–xxxiii, 1–14; M. Cary and E. H. Warmington, *The Ancient Explorers* (London, 1929), pp. 47–52; Hanno, ed. L. Del Turco (Florence, 1958); ed. A. N. Oikonomides (Chicago, 1977); ed. J. Ramin, BAR Suppl. ser. 3 (Oxford, 1976); ed. J. Blomqvist (Lund, 1979).
9 So J. Carcopino, *Le Maroc antique* (Paris, 1943), pp. 73–123, who wondered if the expedition was really a search for gold.
10 Hanno, *Periplus* 17–18. A possible alternative for the 'gorillas' is of very primitive men, like those reported from central Asia in recent times.
11 *NH* vi.199–200.
12 K. G. Sallmann, *Die Geographie des älteren Plinius in ihrem Verhältnis zu Varro* (Berlin and New York, 1971), pp. 68–9.
13 Herodotus iv.42.
14 *Geographi Graeci minores* (n.1), i. 15–96; A. E. Nordenskiöld, *Periplus* (n.1), pp. 6–9.
15 P. Fabre, 'La date de la rédaction du Périple

de Scylax', *Les Etudes Classiques* 1965, 353–66; A. Perotti, *Studi Classici ed Orientali* 1963, 16.
16 The Greek form is βραχειών, perhaps a by-form of βραχίων, either 'shorter' or 'arm'.
17 This assumes an Olympic stade, at 8⅓ stades to the Roman mile.
18 L. Pearson, *The Lost Histories of Alexander the Great*: American Philol. Assoc., Philol. monogr. 20 (New York, 1960), pp. 112–49.
19 *Ind.* 20. 1–2; cf. Pearson, *op. cit.* (n.18), p. 142.
20 Nearchus in Arrian, *Anab.* vi.24.2–4.
21 *NH* vi.96. M. Neubert, 'Die Fahrt Nearchs nach dem konstanten Stadion', *Petermanns Mitteilungen* 74 (1928), 136–43 was on flimsy ground in trying to maintain that Nearchus used a short stade.
22 The story is told by R. Lane Fox, *Alexander the Great* (London, 1973), pp. 405–6 (on p.450 'in Kirman' refers to the province, not town).
23 F. Jacoby, ed., *Die Fragmente der griechischen Historiker* 3C (Leiden, 1958), pp. 592–6.
24 C. F. C. Hawkes, *op. cit.* (n.7).
25 Strabo ii. 4. 1–2.
26 Some have suggested that Pytheas never reached any northerly point. But Geminos, *Introduction to Phaenomena*, vi. 8–9 quotes barbarians showing Pytheas' party where the sun 'went to bed', and certainly took this as proof of a northerly visit.
27 *NH* ii.186–7.
28 Tac. *Agr.* 10.6.
29 *Op. cit.*(n.7), pp. 33–8.
30 *Synt. Math.* ii.6; the *Almagest* was written before the *Geography*, and is thus apt to reflect earlier opinion.
31 C. F. C. Hawkes, *op. cit.* (n.7), pp. 9–10; but the actual latitude of all places in Jutland is too far north, and Pytheas' latitude corresponds rather to Rügen Island, E. Germany.
32 iii. 75–6, perhaps based on Posidonius.
33 Evidently not the P. Crassus who was on Caesar's staff in Gaul, but very likely an ancestor of his.
34 The Scillies may have acted as a trading post. For negative conclusions see Rivet and Smith, *The Place-names of Roman Britain* (London, 1979), p. 43.
35 *Geographi Graeci minores* (n.1), i. 257–305; ed. and transl. by G. W. B. Huntingford (Hakluyt Soc., Ser.II, vol. 151, London, 1980). MSS at Heidelberg, Palatinus graecus 398, 10th century; British Library, Add.MS19391, 14th century.
36 M. G. Raschke, 'New studies in Roman commerce with the East', *ANRW* II.9.2 (1978), 604–1361, esp. 663–5, 754 n.475, 979–90 notes 1342–72; A. Dihle, 'Indienhandel der römischen Kaiserzeit', *ibid.* 570 n.74; G. Mathew, 'The dating and significance of the Periplus of the Erythraean Sea', in H. N. Chittick and R. J. Rotberg, ed., *East Africa and the Orient* (New York, 1975), pp. 154–8, who contrary to the above prefers AD 120–130, advocated by several scholars. For earlier bibliography see A. Diller, *The Tradition of the Minor Greek Geographers* (Lancaster, Pa., 1952), pp. 48–99. The statement of J. A. B. Palmer, '*Periplus maris Erythraei*: the Indian evidence as to the date', *CQ* 41 (1947), 137–40, that Nahapāna's successor Gotamiputra (Gautamiputra) reigned for only about four years from *c.* AD 124 would, if true, allow inscriptions of Nahapāna's reign, dated 41–46 of an unrecorded era, to refer to an era of the Saka dynasty beginning AD 78, as was suggested by A. M. Boyer in 1897. But in fact there are inscriptions dated 18th and 24th years of the reign of Gautamiputra; see Raschke, *op. cit.*, n.475. Leuce Come, mentioned as the start of the road to Petra and as having a small Roman garrison and tax collector, could have been functioning as such, even during the period of the Nabataean kingdom. In addition to various dates under the early Empire, dates between AD 200 and 250 have been suggested: see Raschke, *op. cit.*, n.1343.
37 Erythraean, which gave its name to Eritrea (an Italian neologism) means 'red', and ἐρυθρὴ θάλασσα in Herodotus means the Red Sea. Later it became widened in use to include the Indian Ocean and the Persian Gulf.
38 Excerpts in *Geographi Graeci minores* (n.1), i, liv–lxxiii, 111–95; partly preserved in Diodorus and Photius.
39 *Periplus* 38.
40 *Periplus* 15.
41 The sewing refers to wooden slats built up on a keel hollowed out of a trunk.
42 The word χελώνη is the same as is used above for mountain tortoises, but must here, even without an epithet, refer to turtles.
43 Arrian's most famous work is that known as *Anabasis*, a history of Alexander the Great.
44 *Periplus of the Euxine* 6–7 (*Geographi Graeci minores* i.372–3). This *periplus* dates from *c.* AD 130.
45 *Geographi Graeci minores* (n.1), i. 427–514.
46 *Stadiasmus* 93.
47 Ed. and transl. by J. P. Murphy (Chicago, 1977).
48 *Ora Maritima*, ed. A. Berthelot (Paris, 1934).
49 *Geographi Graeci minores* (n.1), i. 515–62.

50 *Periplus maris exteri* i.1.
51 This name, thought to correspond to Sanskrit *pārasamudra*, 'overseas land', was sometimes attributed to the chief city and one of the rivers rather than to the whole island. In i.35 Marcian, or an interpolator, adds 'and now called Salika' (= Sri Lanka).
52 By Ptolemy's time, and even more by Marcian's, the name Albion was obsolete for Britannia.
53 *Periplus maris exteri* i.22.
54 *Stadiasmus* 57.
55 *Periplus maris exteri* i.4 ff.
56 *Geographi Graeci minores* (n.1), ii. 103–76.
57 *Periegesis* 793–6.
58 *Geographi Graeci minores* (n.1), i. 238–42.
59 *Ibid.* i.196–237.
60 This Scymnus, thought to be the same as the Chian proxenos at Delphi in 185–4 BC, wrote a *Periegesis* in at least 16 books.
61 A. Diller, *The Tradition of the Minor Greek Geographers* (Lancaster, Pa., 1952), pp. 147–64.
62 *Anth. Pal.* ix. 559.

Chapter X

1 G. R. Mair, transl., Callimachus, Lycophron, Aratus (Loeb series, London and Cambridge, Mass., 1921): J. Martin, ed., Florence, 1954).
2 Commagene was a client kingdom in the north of Syria; it became independent in 162 BC.
3 E. L. Stevenson, *Terrestrial and celestial globes* (New Haven, Conn., 1921), i.15 and fig. 7. What appears to be a terrestrial globe is shown on a fresco from a villa at Boscoreale on the slopes of Vesuvius: New York, Metropolitan Museum; Stevenson, *op. cit.*, i.21 and fig. 10.
4 Two great circles which intersect at right angles at the poles and which divide the equinoctial (the circle of the celestial sphere whose plane is perpendicular to the axis of the earth) and the ecliptic (the apparent orbit of the sun) into four equal parts.
5 R. Meiggs, *Roman Ostia* (Oxford, 1960), pl.XVIIIa.
6 *Ibid.* pl.XVIIIb.
7 A. E. M. Johnston, 'The earliest preserved Greek map: a new Ionian coin type', *JHS* 87 (1967), 86–94.
8 Rome, Tittoni collection, late 7th cent. BC: William H. Matthews, *Mazes and Labyrinths* (London, 1922), pp. 157–8, figs. 133–5; G. Q. Giglioli, in *Studi Etruschi* III (1929), 111.
9 Virg. *Aen.* v. 545–603; E. Norden, *Aus altrömischen Priesterbüchern* (Lund and Leipzig, 1939), pp. 188f.
10 *Aen.* v. 588–95, where the young men's movements are compared both to the Cretan labyrinth and to the movements of dolphins.
11 W. H. Matthews, *op. cit.* (n.8), p. 211.
12 *Museo Borbonico* XIV.1852; cf. W. H. Matthews, *op. cit.* (n.8), p. 46, fig. 32; *Enciclopedia di Arte Antica* IV. 437, fig. 507.
13 W. Jobst, *Römische Mosaiken in Salzburg* (Vienna, 1982), pp. 118–23, pls.51–3.
14 J. D. Cowen and I. A. Richmond, 'The Rudge cup', *Archaeologia Aeliana*, 4th ser. XII (1935), 310–42.
15 J. Heurgon, 'The Amiens patera', *JRS* XLI (1951), 22–4.
16 M. Gichon, 'The plan of a Roman camp depicted upon a lamp from Samaria', *Palestine Exploration Quarterly* 104 (1972), 38–58.
17 Livy xli. 28.8–10. Matuta, in origin a dawn goddess, came to be associated with Leucothea, who in Greek mythology was deified as goddess of seas and harbours.
18 Marie-Henriette Quet, *La mosaïque cosmologique de Mérida* (Paris, 1981).
19 G. Gullini, *I mosaici di Palestrina* (Rome, 1956); well illustrated in M. Hadas, *Imperial Rome*: Great Ages of Man (Weert, 1971), pp. 70–71.
20 G. Becatti, *Mosaici e pavimenti marmorei*, in *Scavi di Ostia* IV (Rome, 1961), No. 108, Foro delle Corporazioni, Statio 27, pp. 74ff., tav.CLXXXIV.
21 Dated to *c.* AD 120 by G. Becatti, *op. cit.* (n.20), pp. 42f., No. 64, tav.CVII.
22 Katherine M. D. Dunbabin, *The Mosaics of Roman North Africa* (Oxford, 1978), pp. 57–8, pls. 35–7.
23 E. Kitzinger, 'Studies on late antique and early Byzantine floor mosaics', *Dumbarton Oaks Papers* 6 (1951), 81–122, figs. 18–19.
24 M. Avi-Yonah, *The Madaba Mosaic Map* (Jerusalem, 1954); H. Donner and H. Cuppers, *Die Mosaikkarte von Madeba*: Teil I, Tafelband (Wiesbaden, 1977).
25 (a) Joshua 10.12–14 'stand still, O sun, in Gibeon; stand, moon, in the Vale of Aijalon'; for the topography see Avi-Yonah, *op. cit.* (n.24); (b) Jeremiah 31.15; (c) Genesis 49.13, where the New English Bible makes the tenses present.
26 Phyllis Abrahams, ed., *Les oeuvres poétiques de Baudri de Bourgeuil* (Paris, 1926), pp. 196 ff. Adela, Countess of Blois, was daughter of William the Conqueror. Baudri was Abbot of Bourgeuil 1079–1107, and later Bishop of Dol.

Chapter XI

1 *Claudii Ptolemaei Geographiae Codex Urbinas Graecus 82*, tomus prodromus (Leiden, Leipzig and Turin, 1932), i.108 and elsewhere.
2 'The origin of Ptolemy's Geographia', *Geografiska Annaler* 27 (1945), 318–87.
3 Fischer, *op. cit.* (n.1), i.420–25; *RE* s.v. Pappos (2); Bagrow, *op. cit.* (n.2), esp. 325–7.
4 K. Ziegler s.v. Pappos in *RE* XVIII 2, cols. 1084–1106; id. s.v. Pappos in *Der kleine Pauly*.
5 A. Diller, *The Tradition of the Minor Greek Geographers* (Lancaster, Pa., 1952), p.45, n.104.
6 *Institutiones* i.25.2, ed. R. A. B. Mynors (Oxford, 1937); cf. Angela Codazzi, *Le edizioni quattrocentesche e cinquecentesche della 'Geografia' di Tolomeo* (Milan–Venice, [1950]), p.15.
7 Cassiodorus, *Variae* i.45.
8 H. von Mzik, ed., *Afrika nach der arabischen Bearbeitung der* Γεωγραφικὴ ὑφήγησις *des Claudius Ptolemaeus von Muhammad ibn Musa al-Hwarizmi* (Vienna, 1916); *The Encyclopedia of Islam*, new edn. IV (Leiden, 1973–8), 1077–83, s.v. Kharita.
9 L. arab. cod. Spitta 18, with Nile map fol.29v–30r.
10 H. von Mzik, *op. cit.* (n.8), Taf.1; Youssouf Kamal, *Monumenta Cartographica Africae et Aegypti* III.1 (Leiden, 1930), 525.
11 E. Honigmann, *Die sieben Klimata und die* πόλεις ἐπίσημοι (Heidelberg, 1929), p. 59; L. Bagrow, *op. cit.*(n.2), esp. 320.
12 Algorism, meaning 'arabic notation', is derived from al-Khwarizmi, though not from his co-ordinates.
13 M. Destombes, ed., *Mappemondes A.D. 1200–1500* (Amsterdam, 1964), p. 18; Masudi, *Le livre des prairies d'or* (Paris, 1962), i.76–7. As he could not read Greek, his total is bound to be unreliable.
14 The comparative lists given by Honigmann, *op. cit.* (n.11), pp. 126–7 show the considerable differences between Ptolemy, al-Khwarizmi and al-Battani.
15 A. Nordenskiöld, *Facsimile Atlas* . . ., transl. J. A. Ekelöf and R. Markham (repr. New York, 1973); J. Fischer, *op. cit.* (n.1); B. Harley and D. Woodward, ed., *History of Cartography*, vol. 1 (Chicago, forthcoming).
16 Preserved in the Milan codex Ambrosianus Graecus 43.
17 *Op. cit.* (n.1).
18 The hierarchy of places may not, as claimed by Annalina and M. Levi, *Itineraria picta* (Rome, 1967), p.37, be linked to the Ptolemaic system of 'significant places'.
19 A. L. F. Rivet and Colin Smith, *The Place-names of Roman Britain* (London, 1979), p.139.
20 P. Schnabel, *Text und Karten des Ptolemäus* (Leipzig, 1938).
21 Add. MS. 19391; well reproduced in H. von Mzik, *op. cit.* (n.8), Taf.II.
22 Pencilled note by Leo Bagrow in the British Library catalogue.
23 E. Polaschek, 'Ptolemäus als Geograph', in *RE* Suppl. X, col. 830; J. Fischer, *op. cit.* (n.1), 540–43; J. Babicz, 'The reception of Ptolemy's Geography', in *History of Cartography*, vol. 3 (Chicago, forthcoming).
24 J. Fischer, *op. cit.* (n.1), 205–8.
25 Milan, Ambrosianus F 148, sup.
26 L. Bagrow, 'The maps of Regiomontanus', *Imago Mundi* 4 (1948), 31–2.
27 J. Babicz, 'Donnus Nicolaus Germanus: Probleme seiner Biographie und sein Platz in der Rezeption der ptolemäischen Geographie', in C. Koeman, ed., *Land- und Seekarten im Mittelalter und in der frühen Neuzeit* (Munich, 1980), pp. 1ff.
28 *Geog.* i. 24 *ad fin.*
29 Facsimile with introd. by R. A. Skelton in *Theatrum orbis terrarum*, 1st ser. I (Amsterdam, 1963); Angela Codazzi, *op. cit.* (n.6).
30 Facsimile with introd. by R. A. Skelton in *Theatrum orbis terrarum*, 2nd ser. VI (Amsterdam, 1966).
31 Facsimile with introd. by R. A. Skelton in *Theatrum orbis terrarum*, 1st ser. II (Amsterdam, 1963).
32 A. E. Nordenskiöld, *op. cit.* (n.15), pp. 12, 14.
33 *Geog.* vii. 3 *ad fin.*
34 A. E. Nordenskiöld, *op. cit.* (n.15), p. 27, fig. 14.
35 A. E. Nordenskiöld, *op. cit.* (n.15), p. 72, fig. 40.
36 C. Sanz, *La Geographia de Ptolomeo ampliada con los primeros mapas de América (desde 1507)* (Madrid, 1959).
37 J. N. Wilford, *The Map-makers* (London, 1981), p. 65.
38 F. Laubenberger, 'Ringmann oder Waldseemüller?', *Erdkunde* 13 (1959), 163–79.

Chapter XII

1 Ed. O. Seeck (Berlin, 1876; repr. Frankfurt am Main, 1962); R. Goodburn and P. Bartholomew, ed., *Aspects of the Notitia Dignitatum* (BAR Suppl. Series 15), Oxford, 1976.
2 *Not. Dign.*, *Oc.* iv.5.

3 L. Casson, *Travel in the Ancient World* (London, 1974), p. 315.
4 For bibliography on the emblems see G. Clemente, 'La *Notitia Dignitatum*' (Cagliari, 1968), p. 45, n.40; R. Grigg, 'Inconsistency and lassitude: the shield emblems of the *Notitia Dignitatum*', *JRS* LXXIII (1983), 132–42.
5 There is some doubt whether the province of Valentia existed: see J. G. F. Hind, 'The British provinces of Valentia and Orcades', *Historia* XXIV (1975), 101–11.
6 G. Clemente, *op. cit.* (n.4), pp. 44–56.
7 *Not. Dign., Oc.* i.35; v.131; vii.153, 199.
8 J. J. G. Alexander, 'The illustrated manuscripts of the Notitia Dignitatum', in Goodburn and Bartholomew, *op. cit.* (n. 1), pp. 11–75 and pls. A–C, I–XXIII.
9 The reason why there are two is that Count Ottheinrich was dissatisfied when, instead of being lent the Spirensis, he was given a recent copy. He complained that it was a poor one and was given a second. See I. G. Maier, 'The Barberinus and Munich codices of the "Notitia dignitatum omnium"', *Latomus* 28 (1969), 960–1035.
10 Stephen Johnson, *The Roman Forts of the Saxon Shore* (London, 1976).
11 *Not. Dign., Oc.* xi.60.
12 A. Riese, ed., *Geographi Latini minores* (Heilbronn, 1878, repr. Hildesheim, 1964).
13 Dicuil, *Liber de mensura orbis terrae*, ed. J. J. Tierney: Scriptores Latini Hiberniae 6 (Dublin, 1967).
14 O. Neugebauer, 'A Greek world map', in *Le monde grec: hommages à Claire Préaux* (Brussels, 1975), pp. 312–17, pl. III.
15 Neugebauer writes 'I do not know of a "Marsh" of Meroe'; but λίμνη is a marshy lake, and this surrounded the then existing island of Meroe.
16 The starting point of this section is approximately the end of the Western Empire. The two preceding items may be considered as belonging either to the Eastern Empire or to the early Byzantine Empire.
17 The best is in Florence, Biblioteca Mediceo-Laurenziana IX. 28 (11th century), originally from Mt Athos. Others are Vaticanus graecus 699 (9th century), originally from Constantinople, and Sinaiticus graecus 1186.
18 Ed. Wanda Wolska-Conus (3 vols., Paris, 1968–78).
19 Wanda Wolska, *La topographie Chrétienne de Cosmas Indicopleustès: théologie et science au VIe s.* (Paris, 1962).
20 Wolska, *op. cit.* (n. 19), pl. III a–b.
21 Wolska, *op. cit.* (n. 19), pl. XI.
22 *History of Cartography*, vol. 1 (Chicago, forthcoming).
23 Strabo i. 2. 28.
24 The name seems to have been coined by Gregorio Dati in his poem *La Sfera* (1513).
25 J. T. Lamman, 'The religious symbolism of the T in T-O maps', *Cartographica* 18 (1981), 18–22. The tau cross is mentioned by Isidore (*Etym.* i.3), but not in connection with maps.
26 There is no evidence, despite repeated mentions of 'Orosian maps', that Orosius ever made any maps to accompany his history: see Y. Janvier, *La géographie d'Orose* (Paris, 1982), pp. 59ff.
27 M. A. P. d'Avezac, 'Une digression géographique . . . .', in *Annales des Voyageurs*, ed. V. A. Malto-Brun (Paris, 1872), ii. 193–210; Konrad Miller, *Mappaemundi*, vol. VI (Stuttgart, 1898).
28 W. H. Stahl, 'Astronomy and geography in Macrobius', *Trans. of the Amer. Philol. Assoc.* 73 (1942), 232–58; C. Sanz, *El primer mapa del mundo con la representación de los dos hemisferios, concebido por Macrobio*, Publicaciones de la Real Soc. Geog., Serie B, No. 455 (Madrid, 1966).
29 Ed. M. Pinder and G. Parthey, Ravennatis anonymi *Cosmographia* et Guidonis *Geographica* (Berlin, 1860); J. Schnetz, *Untersuchungen über die Quellen der Kosmographie des anonymen Geographen von Ravenna* (Munich, 1942).
30 Konrad Miller, *op. cit.* (n.27), Tafel 1.
31 Konrad Miller, *Itineraria Romana* (Stuttgart, 1916), pp. xxvii–xxix.
32 A. L. F. Rivet and Colin Smith, *The Place-names of Roman Britain* (London, 1979), p. 359, col. 2.
33 It is most unlikely that this El- reflects a non-Latin origin; E and F are very similar in rustic capitals.
34 W. Horn and E. Born, *The Plan of St Gall*, 3 vols. (Berkeley, California, 1979); P. D. A. Harvey, *The History of Topographical Maps* (London, 1980), pp. 131–2.
35 Konrad Miller, *Mappae Arabicae* i.1 (Stuttgart, 1926).
36 *Ibid.* i.2 (Stuttgart, 1926).
37 Roger's name for the work was Nuzhat al-Mushtāk ('the devotee's journey'); Idrisi called it Kitāb Rudjān, 'Roger's book'.
38 E. Honigmann, *Die sieben Klimata und die* πόλεις ἐπίσημοι (Heidelberg, 1929), pp. 182–3 with references. The criticism goes back indirectly to an Arabic encyclopaedia of 1348.
39 E.g. in the Oxford copy, Bodleian Uri MS 887, reproduced (upside down) in L. Bagrow,

rev. R. A. Skelton, *History of Cartography* (London, 1964), p.57.

40 Honigmann, *op. cit.* (n. 38), *passim.*

41 ix.1.2.

42 Honigmann, *op. cit.*, pp. 10ff., speaks of strips of land running east–west, but there is no real evidence that they were anything other than particular latitudes.

43 *NH* vi. 211–20.

44 M. Destombes, ed., *Mappemondes A.D. 1200–1500* (Amsterdam, 1964); David Woodward in *History of Cartography*, vol. 1 (Chicago, forthcoming).

45 Alternatively *estoire* can of course refer to a book; but although that has been suggested in this context, it does not suit the following words, *ki lat fet e compasse* (= qui l'a fait et compassé), referring to Richard of Haldingham (see below).

46 G. R. Crone, *The Hereford World Map* (London, 1948).

47 As a Latin word, *ornesta* is meaningless. It has reasonably been supposed to have arisen from an acronym, the first component being an abbreviation of *Orosii.*

48 Luke 2.1.

49 The commonest number of winds in the Classical period is twelve, and early medieval compass cards are circles with twelve directional points; later ones tend to have 32.

50 G. R. Crone, 'New light on the Hereford Map', *Geographical J.* 131 (1965), 447–62.

51 E. F. Jomard, *Les monuments de la géographie* (Paris, 1843–6), pls.v.1–3 (39–41); MSS in Corpus Christi College, Cambridge, and in the British Library (Royal 14C. VII, f.1b).

52 L. Bagrow, rev. R. A. Skelton, *History of Cartography* (Cambridge, Mass and London, 1964), pl.XXXII.

53 *Ibid.* pl.XXXV.

54 Bibl. Nat., Espagnol 30; G. Grosjean, ed., *Mapamundi: the Catalan Atlas of the Year 1375* (Dietikon–Zurich, 1978).

55 *Ibid.* pp. 18–19, fig. 4.

56 G. Grosjean, ed., *Der Catalane Weltatlas vom Jahre 1375* (Zurich, 1977), p.16 argues that the medieval portolan tradition is at least in part derived from the maps of the Agrimensores. But the latter were land maps whereas the portolans were essentially sea charts, and his arguments seem tenuous.

57 Bagrow, rev. Skelton, *op. cit.* (n.52), pl.XLI.

SELECT BIBLIOGRAPHY

ABBREVIATIONS

ANRW	*Aufstieg und Niedergang der römischen Welt*, ed. Hildegard Temporini et al.
BAR	British Archaeological Reports
JHS	*Journal of Hellenic Studies*
RE	Pauly-Wissowa, *Realenzyklopädie der klassischen Altertumswissenschaft*

Aristotle, *Meteorologica*, ed. H. D. P. Lee (London, 1952).

Aujac, Germaine, *Strabon et la science de son temps* (Paris, 1966).

Babicz, J., 'The Reception of Ptolemy's *Geography*', in *The History of Cartography*, Vol.3 (Chicago and London, forthcoming).

Bagrow, L., *History of Cartography*, rev. R. A. Skelton (London, 1964).

Bagrow, L., 'The Origin of Ptolemy's Geographia', *Geografiska Annaler* 27 (1945), 318–87.

Brendel, O. J., *Symbolism of the Sphere* (Leiden, 1977).

Brown, Lloyd A., *The Story of Maps* (Boston, Mass., 1949).

Bunbury, Sir E. H., *History of Ancient Geography*, 2 vols. (London, 1883); repr. with introd., by W. H. Stahl (New York, 1959).

Butzmann, H., ed., *Corpus Agrimensorum: Codex Arcerianus A der Herzog-August Bibliothek* (Leiden, 1970).

Carder, J. N., *Art historical Problems of a Roman Survey Manuscript: the Codex Arcerianus A* (New York and London, 1978).

Carettoni, G. et al., *La pianta marmorea di Roma antica*, 2 vols. (Rome, 1960).

Cary, M. and Warmington, E. H., *The ancient Explorers* (London, 1929).

Cebrian, K., *Geschichte der Kartographie*, vol. 1 (no more pubished; Gotha, 1922).

Corpus Agrimensorum: Blume, F. et al., ed., *Die Schriften der römischen Feldmesser*, 2 vols. (Berlin, 1848–52).

Corpus agrimensorum Romanorum I, ed. C. O. Thulin (no more published; rev. edn., Stuttgart, 1971).

Cosmas Indicopleustes, ed. Wanda Wolska-Conus, 3 vols. (Paris, 1968–78).

Cumont, F., 'Fragment de bouclier portant une liste d'étapes', *Syria* VI (1925), 1–15 and pl.I.

Cuntz, O., ed., *Itineraria Romana* I (Leipzig: Teubner, 1929).

Destombes, M., *Mappemondes, A.D. 1200–1500* (Amsterdam, 1964).

Dilke, O. A. W., 'Illustrations from Roman Surveyors' Manuals', *Imago Mundi* XXI (1967), 9–29.

Dilke, O. A. W., 'Maps in the Treatises of Roman Land Surveyors', *The Geographical J.* CXXVII (1961), 417–26.

Dilke, O. A. W., *The Roman Land Surveyors: an Introduction to the Agrimensores* (Newton Abbot, 1971); Ital. transl. by Gabriella Ciaffi Taddei, *Gli Agrimensori di Roma antica* (Bologna, 1979).

Diller, A., *The Tradition of the minor Greek Geographers* (Lancaster, Pa., 1952).

Donner, H. and Cuppers, H., *Die Mosaikkarte von Madeba* I (Wiesbaden, 1977).

Geographi Graeci minores, ed. C. Müller, 2 vols. (Paris, 1832).

Geographi Latini minores, ed. A. Riese (Heilbron, 1878).

Goodburn, R. and Bartholomew, P., ed., *Aspects of the Notitia Dignitatum*. BAR Suppl. Ser. 15 (Oxford, 1976).

Harley, J. B. and Woodward, D., ed., *The History of Cartography*, Vol. 1 (Chicago and London, forthcoming).

Harvey, P. D. A., 'Cadastral, urban and topographical Cartography', in *The History of Cartography*, Vol. 1, Sect. D (The Cartographic Traditions of Medieval Europe) (Chicago and London, forthcoming).

Harvey, P. D. A., *The History of Topographical Maps* (London, 1980).

Hawkes, C. F. C., *Pytheas: Europe and the Greek Explorers* (Oxford, 1975).
Heidel, W. A., *The Frame of the ancient Greek Maps* (New York, 1937).
Herodotus, *Histories, passim.*
Hipparchus, *The Geographical Fragments*, ed. D. R. Dicks (London, 1960).
Honigmann, E., *Die sieben Klimata und die* ΠΟΛΕΙΣ ΕΠΙΣΗΜΟΙ (Heidelberg, 1929).
Kish, G., ed., *A Source Book in Geography* (Cambridge, Mass., 1973).
Klotz, A., 'Die geographischen commentarii des Agrippa und ihre Überreste', *Klio* 24 (1931), 38–58, 386–466.
Kubitschek, W., 'Itinerarien', in *RE* XVIII. Halbband (1966), cols. 2308–63.
Levi, Annalina and M., *Itineraria picta: Contributo allo studio della Tabula Peutingeriana* (Rome, 1967).
Macrobius, *Commentary on the Dream of Scipio*, transl. W. H. Stahl (New York, 1952). Latin text ed. J. Willis (Leipzig: Teubner, 1963).
Mela, Pomponius, *Chorographia*, ed. G. Ranstrand (Göteborg, 1971); ed. P. G. Parroni (Rome, 1984).
Mette, H. J., *Sphairopoiia, Untersuchungen zur Kosmologie des Krates von Pergamon* (Munich, 1936).
Millard, A., 'Mesopotamia', in *The History of Cartography*, Vol.1 (Chicago and London, forthcoming).
Miller, Konrad, *Itineraria Romana* (Stuttgart, 1916).
Miller, Konrad, *Mappaemundi: Die älteste Weltkarten*, 6 vols. (Stuttgart, 1895–8).
Miller, Konrad, *Die Peutingersche Tafel*, repr. based on the above (Stuttgart, 1962).
Neugebauer, O., *A History of Ancient Mathematical Astronomy*, 3 vols. (Berlin, Heidelberg and New York, 1975).
Newton, R. R., *The Crime of Claudius Ptolemy* (Baltimore, 1977).
Nordenskiöld, A. E., *Facsimile-Atlas to the early History of Cartography*, transl. J. A. Ekelöf and C. R. Markham (Stockholm, 1889).
Nordenskiöld, A. E., *Periplus*, transl. F. A. Bather (Stockholm, 1897).
Notitia Dignitatum, ed. O. Seeck (Berlin, 1876).
Piganiol, A., *Les documents cadastraux de la colonie romaine d'Orange: Gallia* Suppl. 16 (Paris, 1962).
Pliny, *Natural History* III–VII, transl. H. Rackham (Loeb Classical Library, Cambridge, Mass., and London, 1969). Also individual books in the Budé series (continuing).
Polaschek, E., 'Karten', in *RE* XX. Halbband (1919), cols. 2022–2149.
Polaschek, E., 'Ptolemaios als Geograph', *RE* Suppl. 10 (1965), cols. 680–833.
Polaschek, E., 'Ptolemy's Geography in a new Light', *Imago Mundi* 14 (1959), 17–37.
Ptolemy, *Almagest*, ed. J. L. Heiberg, 2 vols. (Leipzig: Teubner, 1898, 1903).
Ptolemy, *Almagest*, transl. C. J. Toomer (London, 1984).
Ptolemy, *Geographia*, ed. C. Müller, 2 vols. and Tabulae (Paris, 1883–1902: text of Books I–VI only).
Ptolemy, *Geographia*, ed. C. Nobbe (Leipzig, 1843–5; repr. with introd. by A. Diller, Hildesheim, 1966).
Ptolemy, *Die Geographie des Ptolemaeus: Galliae Germania Rhaetia Noricum Pannoniae Illyricum Italia*, ed. O. Cuntz (Berlin, 1923).
Ptolemy, *The Geography of Claudius Ptolemy*, transl. E. L. Stevenson (New York, 1922).
Pytheas von Massalia, *Fragmenta*, ed. H. J. Mette (Berlin, 1952).
Ravennatis anonymi *Cosmographia* et Guidonis *Geographica* (Berlin, 1860).
Reed, N., 'Pattern and Purpose in the Antonine Itinerary', *American J. of Philology* 99 (1978), 228–54.
Richmond, I. A. and Crawford, O. G. S., 'The British Section of the Ravenna Cosmography', *Archaeologia* XCIII (1949), 1–50.
Rivet, A. L. F., 'The British Section of the Antonine Itinerary', *Britannia* 1 (1970), 34–68.
Rivet, A. L. F. and Smith, C., *The Place-names of Roman Britain* (London, 1979).
Rodriguez Almeida, E., *Forma Urbis marmorea: Aggiornamento generale 1980* (Rome, 1981).
Sallmann, K. G., *Die Geographie des älteren Plinius in ihrem Verhältnis zu Varro* (Berlin and New York, 1971).
Salviat, F., 'Orientation, extension et chronologie des plans cadastraux d'Orange', *Revue Archéologique de Narbonnaise* X (1977), 107–18.
Schmitt, P., 'Recherches des règles de construction de la cartographie de Ptolémée', *Colloque international sur la cartographie archéologique et historique, à la mémoire de F. Oudot de Dainville*, ed. R. Chevallier (Tours, 1976), pp. 27–61.
Schnabel, P., *Text und Karten des Ptolemäus* (Leipzig, 1938).
Schnetz, J., *Untersuchungen über die Quellen der Kosmographie des anonymen Geographen von Ravenna* (SB der Bayerischen Akad. d.

Wissenschaften, Phil.-hist. Abt., Heft 6: Munich, 1942).
Schütte, G., *Ptolemy's Map of Northern Europe: a Reconstruction of the Prototypes* (Copenhagen, 1917).
Sherk, R. K., 'Roman geographical Exploration and military Maps', *ANRW* II.1 (Berlin, 1974), 534–60.
Shore, A. F., 'Ancient Egypt', in *The History of Cartography*, Vol. 1 (Chicago and London, forthcoming).
Skelton, R. A., Facsimiles of early printed editions of Ptolemy's *Geography* ('Cosmography') in *Theatrum orbis terrarum* (Amsterdam, 1963–6).
Stahl, W. H., *Ptolemy's Geography: A Select Bibliography* (New York, 1953).
Stevenson, E. L., *Terrestrial and Celestial Globes. . .*, 2 vols. (New Haven, Conn., 1921).
Strabo, *Géographie*, ed. Germaine Aujac et al. (Budé series, Paris, from 1966).
Strabo, *The Geography*, transl. H. L. Jones (Loeb Classical Library, London, 1917–32).
Thomson, J. O., *History of Ancient Geography* (Cambridge, 1918).
Tozer, H. F., *A History of Ancient Geography*, 2nd edn. (Cambridge, 1935).
Tudeer, L. O. T., 'On the Origin of the Maps attached to Ptolemy's Geography', *JHS* XXXVII (1917), 62–76.
Warmington, E. H., *Greek Geography* (London, 1934).
Weber, E., *Tabula Peutingeriana: Codex Vindobonensis 324* (Graz, 1976).
Wilford, J. N., *The Mapmakers* (London, 1981).
Woodward, D., 'Medieval Mappaemundi', in *The History of Cartography*, Vol. 1 (Chicago and London, forthcoming).
Zicàri, I., 'L'anemoscopio Boscovich del Museo Oliveriano di Pesaro', *Studia Oliveriana* II (Pesaro, 1954), 69–75.

SOURCES OF ILLUSTRATIONS

Alinari 7; Amman Archaeological Museum 25; Arthaud 3; Centro Camuno di Studi Preistorici, Capo di Ponte Fig. 18; Hereford, by permission of the Dean and Chapter of Hereford Cathedral 31; Hirmer Fotoarchiv 4; Jena, Friedrich Schiller Universität 15; London, British Library 26, 32; British Museum 2; Oxford, Museum of the History of Science 6; Pesaro, Museo Oliveriano 5, 21; Philadelphia, University of Pennsylvania Museum 1; Rome, Museo di Roma 18; St Gallen, Stiftsbibliothek 30; Vatican, Biblioteca Apostolica Vaticana 9, 10, 11, 14, 16, 19, 27, 28; Venice, Biblioteca Marciana 29; Vienna, Österreichische Nationalbibliothek 22, 23; Wolfenbüttel, Herzog August Bibliothek 12, 13.

Fig. 1 from J. Ball, *Egypt in the Classical Geographers* (Cairo, 1942)
Pl. 24 from G. Becatti, *Scavi di Ostia*, Vol. IV (Rome, 1953)
Fig. 25 from R. Chevallier, *Les Voies Romaines* (Paris, 1972)
Fig. 16 from the author's *Gli Agrimensori di Roma antica* (Bologna, 1979), and some others from his *The Roman Land Surveyors* (Newton Abbot, 1971)
Figs. 6, 11 and 23 were drawn by John Dixon
Fig. 27 after M. Gichon, 'The Plan of a Roman Camp depicted upon a lamp from Samaria', *Palestine Exploration Quarterly*, Vol. 104 (1972)
Fig. 3 after H. D. P. Lee, *Aristotle's Meteorologica* (London, 1952)
Fig. 22 after Annalina and M. Levi, *Itineraria picta: contributo allo studio della Tabula Peutingeriana* (Rome, 1967)
Pl. 17 from J. Mellaart, *Çatal Hüyük* (London, 1967)
Pl. 8 from E. Nowotny, *Römische Forschung in Österreich 1912–24* (Berlin, 1926)
Figs. 9 and 10 after Pauly, *Real-Encyclopädie der Classischen Altertumswissenschaft*, Suppl.X (Stuttgart, 1965)
Pl. 20, Figs. 20 and 30 from A. Piganiol, *Les documents cadastraux de la colonie romaine d'Orange* (Paris, 1962)
Fig. 29 after O. Seeck, *Notitia Dignitatum* (Rome, 1876)

INDEX